Complex Interactions in Lake Communities

Stephen R. Carpenter
Editor

Complex Interactions in Lake Communities

With 51 Figures

Springer-Verlag
New York Berlin Heidelberg
London Paris Tokyo

Stephen R. Carpenter
Department of Biological Sciences
University of Notre Dame
Notre Dame, Indiana 46556-0369
U.S.A.

Library of Congress Cataloging-in-Publication-Data
Complex interactions in lake communities / [edited by] Stephen R.
  Carpenter.
    p. cm.
    Proceedings of a workshop held at the University of Notre Dame,
  March 21–26, 1987.
    Includes index.
    1. Lake ecology—Congresses.   2. Biotic communities—Congresses.
  I. Carpenter, Stephen R.
  QH541.5.L3C66 1988
  574.5'26322—dc19   87-37625

QH
541.5
.L3
C66
1988

Typeset by David E. Seham Associates, Inc., Metuchen, New Jersey.
Printed and bound by Edwards Brothers, Inc., Ann Arbor, Michigan
Printed in the United States of America.

9 8 7 6 5 4 3 2

ISBN 0-387-96684-6 Springer-Verlag New York Berlin Heidelberg
ISBN 3-540-96684-6 Springer-Verlag Berlin Heidelberg New York

Foreword

In its statutory authority (National Science Foundation Act of 1950, as amended), the NSF is directed to both initiate and support basic scientific research. In its Ecology Program, one mode of initiating research is to encourage the development of new ideas through advisory workshops. The NSF is specifically directed to strengthen our nation's research potential. In addition, stimulating new approaches to research will continue to be prominent in the coming years as federal attention is given to increasing the innovativeness and competitiveness of the U.S. in science and engineering.

A decision to initiate a workshop does not arise de novo in the Ecology Program. Rather, it emerges from panel discussions, conversations with investigators at meetings or on the phone, and from discussions between program officers in the Division of Biotic Systems and Resources.

This workshop was developed to provide advice to the NSF and the limnological community. Some NSF perceptions on future funding for ecological research on lake communities are presented here. Researchers often mentioned a paucity of innovative lake ecology at the community level. This perception was accompanied by a certain frustration since lakes probably have the best empirical data base of any natural environment and should continue to lead in the development of ecological concepts. Members of NSF advisory panels sometimes expressed similar concerns during consideration of proposals for lake research. Also, the planning/budgeting process for the Ecology Program suggested that lake community ecology needed some rethinking and vitalization. Program personnel found it difficult to identify research on lakes as sufficiently innovative and general to justify arguing for budgetary enhancements to the Program in various budget and planning documents. This is not to say that excellent science was not being done by limnologists. However, good science on well-studied subjects is not exceptional within the NSF, nor is it very impressive when the NSF requests and must justify budget increases for ecology studies. In addition, there was a dearth of competitive proposals being submitted to the Ecology Program on limnological subjects, generally, and on aquatic community ecology, specifically.

Several concerns like these led Ecology Program personnel to stimulate this workshop to rethink research on the community ecology of lakes. When approached, Dr. Stephen Carpenter agreed to take the lead and subsequently invested a great deal of time in organizing this meeting, proposing it formally to the NSF for external review, responding to concerns expressed by peer reviewers and NSF staff, administering the actual conference, and engendering and editing this volume. The outstanding contributions to this reference text, both syntheses of emerging concepts and scenarios for future research, convince us that the product is very worthwhile and important.

The NSF expects this book to help set the tone of community level research in lakes for the next 5 to 10 years by outlining concepts and setting realistic priorities for future funding. Neither the Foundation nor the organizers of this workshop are advocating any particular approach to limnology at the expense of other types of research. However, the NSF does seek projects that will generate the greatest conceptual advances for the limited funding available.

Research funding will continue to be highly competitive both for individual investigators and among subdisciplines within community ecology. Even with the proposed doubling of NSF's budget by 1992, the rate of funding for proposals will probably not climb substantially since both the number of fundable projects and their budget size are expected to continue increasing over the 1988 to 1992 period. Peer reviewers, advisory panelists, and NSF program officers will set funding priorities using the SOS criteria of scientific merit—soundness, originality, and significance. Thus, ideas presented in this volume should directly improve the caliber of future proposals, and should help to increase the competitiveness of aquatic studies.

The perceived need for larger-scale collaborative studies should not discourage individual researchers from pursuing their chosen avenues of inquiry. If a single-investigator project is scientifically original and significant, it will also be given high priority for support. However, in many subdisciplines of ecology, 3-year projects with three or more collaborators are increasingly successful, and there is Foundationwide support for larger projects with larger budgets, more collaboration, and the development of research centers for specific research activities. These thrusts are prominent in the NSF budget document submitted to Congress for fiscal year 1988, and will persist, as NSF's goal is to improve U.S. innovation and competitiveness in science.

There is no stigma as far as the NSF is concerned in studying theoretical questions using manipulated or man-made systems. Interest in this aspect of ecology was visible also in the participants and topics included in the joint meeting of the Ecological Society of America and the IVth International Congress of Ecology, Syracuse, New York (August 1986). Basic research is being conducted in agroecosystems, and also in reservoirs and regulated rivers by aquatic ecologists. When novel ecological research can be performed in systems controlled by man, investigators should not hesitate to propose such research for NSF support. Increasingly, data sets developed for applied purposes such as water quality evaluation contribute answers to basic ecological questions about aquatic systems. This trend is to be encouraged.

There are many exciting opportunities for new research on complex interactions occurring in the biotic communities of lakes, and between them and their surrounding landscapes. It is our hope that the concepts and priorities given in this book will act as catalysts, leading to a new era of limnological research.

On behalf of the Foundation, we thank Dr. Stephen Carpenter, members of the Steering Committee, and workshop participants for developing their ideas and hypotheses into the reality of this volume. We also thank the University of Notre Dame for its financial and logistic contributions to this effort.

Garth W. Redfield
Patrick W. Flanagan
National Science Foundation

Preface

From March 21 to 26, 1987, a group of 40 ecologists met at the University of Notre Dame to discuss complex interactions in lake communities. This volume emanated from that workshop. The goals of the workshop and this book are to stimulate creativity and enhance learning rates in aquatic ecology. To these ends, we have focused on certain frontiers in aquatic community ecology which encompass major unsolved problems and likely growth points for progress.

Community processes were the mandated focus of the workshop. Species interactions are therefore the central concern of this volume. However, community processes are best understood by reference to both higher and lower levels. This elementary lesson from hierarchy theory accords with long-standing traditions in community ecology, and prompted us to consider ecosystem processes (biogeochemical cycles, trophic status) and autecological properties (life histories, ecophysiologies) as contexts and constraints of species interactions.

The book leads the reader to touchstone papers and key concepts, but we did not attempt a thorough historical review of lake community ecology. Since the book reflects the opinions of scientists experienced with many different taxocenes, lake types, and methodologies, we do not advocate any single approach to limnological research. Rather, we seek a perspective that allows us to determine efficient, appropriately scaled research plans for a wide range of limnological problems. I regard this volume as a prospectus for progress. If it is not obsolete in 10 years, then we have failed in our purpose for publishing it.

The workshop and this book developed through the contributions of many people. I thank Garth Redfield for his enthusiasm and helpful advice throughout the project. Patrick Flanagan and John Brooks provided valuable suggestions prior to the workshop and during preparation of this volume. Jim Kitchell contributed to the project from its inception and in many ways. He and the other members of the Steering Committee, Bill Neill, Don Scavia, and Earl Werner, were instrumental in planning the scientific discussions at the workshop and summarizing the results of the workshop. Discussion leaders and rapporteurs worked very hard to produce timely group reports (Chapters 10–14, 16). I thank them for their professionalism and for their forbearance under persistent urging to meet short deadlines. Without the generosity, creativity, and collegiality of the workshop participants, this project could not have succeeded. I thank them all for their contributions.

Harriet Baldwin of Notre Dame's Center for Continuing Education superbly handled all logistics of the workshop. Loretta Wasmuth and her assistants Joan Smith and Victoria Harman, at Notre Dame's Department of Biological Sciences, ably converted the manuscript to standard computer files in order to facilitate the typesetting. The manuscript was completed during a sabbatical at the Center for Limnology at the University of Wisconsin–Madison, which was supported by the National Science Foundation (BSR-86-06271), the University of Notre Dame, and the W. V. Kaeser Visiting Scholar fund of the CFL. I thank John Magnuson and the CFL staff for their hospitality and assistance.

I am extremely grateful to the Ecology Program of the NSF for supporting the workshop and writing of this book through grant BSR-86-18434. Finally, I thank the University of Notre Dame for its many contributions to the workshop.

Madison, Wisconsin   Stephen R. Carpenter
January 1988

Contents

Multiple Causality and Temporal Pattern in Lake Ecosystems

Reports from Group Discussions

Synthesis

List of Participants and Contributors

Workshop Participants

Timothy F.H. Allen, Department of Botany, University of Wisconsin, Madison, WI 53706, U.S.A.

John W. Barko, Environmental Laboratory, Waterways Experiment Station, Vicksburg, MS 39180, U.S.A.

Steven M. Bartell, Environmental Sciences Division, Oak Ridge National Laboratory, Oak Ridge, TN 37831, U.S.A.

Stephen R. Carpenter, Department of Biological Sciences, University of Notre Dame, Notre Dame, IN 46556-0369, U.S.A.

Larry B. Crowder, Department of Zoology, North Carolina State University, Raleigh, NC 27695, U.S.A.

Donald L. DeAngelis, Environmental Sciences Division, Oak Ridge National Laboratory, Oak Ridge, TN 37831, U.S.A.

Stanley I. Dodson, Department of Zoology, University of Wisconsin, Madison, WI 53706, U.S.A.

Ray W. Drenner, Biology Department, Texas Christian University, Fort Worth, TX 76129, U.S.A.

Thomas M. Frost, Center for Limnology, University of Wisconsin, Madison, WI 53706, U.S.A.

Donald J. Hall, Department of Zoology, Michigan State University, East Lansing, MI 48824, U.S.A.

Robert W. Howarth, Section of Ecology and Systematics, Cornell University, Ithaca, NY 14853, U.S.A.

Stuart H. Hurlbert, Department of Biology, San Diego State University, San Diego, CA 92182, U.S.A.

W. Charles Kerfoot, Great Lakes Research Division, University of Michigan, Ann Arbor, MI 48109, U.S.A.

James F. Kitchell, Center for Limnology, University of Wisconsin, Madison, WI 53706, U.S.A.

David M. Lodge, Department of Biological Sciences, University of Notre Dame, Notre Dame, IN 46556-0369, U.S.A.

Donald J. McQueen, Biology Department, York University, Toronto M3J 1P3, Canada

John M. Melack, Department of Biological Sciences, University of California, Santa Barbara, CA 93106, U.S.A.

Bruce Menge, Department of Zoology, Oregon State University, Corvallis, OR 97331, U.S.A.

Edward L. Mills, Cornell Biological Field Station, Bridgeport, NY 13030, U.S.A.

Gary G. Mittelbach, Department of Zoology, Ohio State University, Columbus, OH 43210, U.S.A.

William E. Neill, Department of Zoology, University of British Columbia, Vancouver, BC V6T 2A9, Canada

Michael Pace, Institute of Ecosystems Studies—NYBG, Millbrook, NY 12545, U.S.A.

Hans Paerl, Institute of Marine Science, University of North Carolina, Morehead City, NC 28557, U.S.A.

Lennart Persson, Fish Ecology Research Group, Department of Ecology— Limnology, S-221 00 Lund, Sweden

Karen G. Porter, Zoology Department, University of Georgia, Athens, GA 30602, U.S.A.

John Priscu, Department of Biology, Montana State University, Bozeman, MT 59717, U.S.A.

Garth Redfield, Ecology Program, National Science Foundation, Washington, DC 20550, U.S.A.

Bo Riemann, Freshwater Biological Laboratory, University of Copenhagen, DK-3400 Hillerod, Denmark

Craig D. Sandgren, Department of Biology, University of Wisconsin, Milwaukee, WI 53201, U.S.A.

Donald Scavia, Great Lakes Environmental Research Laboratory, National Oceanic and Atmospheric Administration, Ann Arbor, MI 48105, U.S.A.

Ulrich Sommer, Max-Planck-Institut für Limnologie, D-2320 Plön, Federal Republic of Germany

Craig N. Spencer, Flathead Lake Biological Station, University of Montana, Polson, MT 59860, U.S.A.

W. Gary Sprules, Department of Zoology, University of Toronto, Mississauga, Ontario L5L 1C6, Canada

Roy A. Stein, Department of Zoology, Ohio State University, Columbus, OH 43210-1293, U.S.A.

John Stockner, Canada Fisheries and Oceans, West Vancouver Laboratory, West Vancouver, BC V7V 1N6, Canada

David Strayer, Institute of Ecosystem Studies—NYBG, Millbrook, NY 12545, U.S.A.

Stephen T. Threlkeld, Biological Station, University of Oklahoma, Kingston, OK 73439, U.S.A.

John E. Titus, Department of Biological Sciences, State University of New York, Binghamton, NY 13901, U.S.A.

Michael J. Vanni, Center for Limnology, University of Wisconsin, Madison, WI 53706, U.S.A.

Earl E. Werner, Division of Biological Sciences, University of Michigan, Ann Arbor, MI 48109, U.S.A.

Additional Contributors

Gunnar Andersson, Fish Ecology Research Group, Department of Ecology—Limnology, S-221 00 Lund, Sweden

Antoinette L. Brenkert, Science Applications International Corporation, Oak Ridge, TN 37830, U.S.A.

Gary L. Fahnenstiel, Great Lakes Environmental Research Laboratory, National Oceanic and Atmospheric Administration, Ann Arbor, MI 48105, U.S.A.

John L. Forney, Department of Natural Resources, Cornell University, Ithaca, NY 14583, U.S.A.

Robert H. Gardner, Environmental Sciences Division, Oak Ridge National Laboratory, Oak Ridge, TN 37831, U.S.A.

Stellan F. Hamrin, Fish Ecology Research Group, Department of Ecology—Limnology, S-221 00 Lund, Sweden

Robert Hodson, Department of Microbiology, University of Georgia, Athens, GA 30602, U.S.A.

Lars Johansson, Fish Ecology Research Group, Department of Ecology—Limnology, S-221 00 Lund, Sweden

Robert V. O'Neill, Environmental Sciences Division, Oak Ridge National Laboratory, Oak Ridge, TN 37831, U.S.A.

CHAPTER 1

Introduction

Stephen R. Carpenter and James F. Kitchell

The magnitude of temporal variability and the frequency of counterintuitive behavior evidenced by lake communities continue to surprise limnologists and defy prediction. Over the past 2 decades, paradigms based upon nutrient loading, organic energy transfer, or predation have developed, and each has successfully answered important questions. Taken alone, each of these paradigms explains certain aspects of lake community dynamics but seems unable to encompass the full range of community variability. Some of these insufficiencies may be resolved by determining the proper hierarchical relationship of key mechanisms. For example, nutrient loading and organic energy transfer establish potential lake productivity, while food web processes determine the realization of that potential (Carpenter et al. 1985). On the other hand, a new synthesis may come from a completely novel view of lakes rather than a hierarchical fusion of existing paradigms. Regardless of its origins, a more comprehensive theory of lake community dynamics is necessary, and must adopt a broader perspective than any one of the existing paradigms of lake ecology. We suggest that analyses of complex interactions, which involve elements of each of the existing paradigms, could lead to a new and more comprehensive perspective. This book takes a step toward a new synthesis by outlining a program of research on complex interactions.

In addition to the scientific issues elaborated below, we have a more practical purpose in publishing this book. "Complex interactions" is a unifying theme that provides a common ground for the diverse concerns of aquatic ecology and focuses the attention of limnologists on general questions of broad significance. Many aquatic ecologists are concerned about the fragmentation of the discipline and the fact that it is losing ground in the funding competition with other sciences, including other branches of ecology. This concern is evident in the preface by Redfield and Flanagan, has been expressed by NSF panel members, and was manifest at the Complex Interactions Workshop. Kitchell et al. (this volume) suggest means for addressing some of these issues, based on discussions of workshop participants.

What Are Complex Interactions?

Complex interactions result from multiple pathways linking organisms and/ or abiotic resources such as nutrients or detritus. Links incorporate transfers of organisms, energy, nutrients, and/or information (e.g., chemical signals, visual signals, behavioral stimulus–response mechanisms, etc.). Interactions resulting from networks of such links are complex because they involve indirect effects and time lags. Investigators often find that neglected components of the system are important. Surprising community dynamics arise from temporal and spatial heterogeneity in the strength or existence of key links.

Established ways of linking the components of aquatic communities, such as biogeochemical cycles or food webs, are types of complex interactions. Indirect effects resulting from chained food web links (Diamond and Case 1986, Kerfoot and Sih 1987) represent another kind of complex interaction of great current interest to ecologists. We have defined complex interactions as broadly as possible to include: transfers of more than one kind of material and/or information; links that may exist irregularly, seasonally, or continuously; and spatial patterns of links that may be homogeneous or heterogeneous. This inclusive definition is appropriate to the breadth and diversity of the questions raised by this volume.

Resource-based competition theory for phytoplankton provides a straightforward and thoroughly developed example of a complex interaction (Tilman et al. 1982). According to this theory, phytoplankton composition is the result of nutrient supply ratios moderated by the algae themselves through a matrix of links among species populations and individual nutrient pools. The theory combines biogeochemical and population processes. Indirect effects are involved because competitors affect each other through their effects on shared resource pools. Tilman et al. (1982) note the need to broaden the theory by incorporating physical factors (such as temperature) and the effects of grazers and " . . . suggest that a synthesis of approaches used by population ecologists and ecosystem ecologists is needed . . . " (p. 367).

Numerous examples show that progress in understanding of important ecological questions can result from recognition of complex interactions. Consider the unexpected increase in *Daphnia* density in Lake Washington, which increased the transparency of the lake (Edmondson 1979; Edmondson and Litt 1982). In explaining the dynamics of *Daphnia* and algae in Lake Washington, Edmondson and his coworkers extended their frame of reference to include food web processes as well as the reductions in phosphorus input that had caused earlier changes in the lake. Progress in limnology has often come from changes in perspective following surprising results (see Neill, this volume; Kitchell et al., this volume).

Another example of progress through a change in perspective is the recognition by R. G. Wetzel and his coworkers that explanations of the pelagic carbon cycle and long-term succession of Lawrence Lake required a thorough analysis of the littoral zone and its biogeochemical links with the pelagic zone (Wetzel and Allen 1970; Wetzel et al. 1972; Wetzel 1979). Wetzel was among the first to integrate littoral processes with more conventional pelagic limnology through complex biogeochemical interactions involving several elements and time scales. These processes are complemented by a host of interspecific interactions both within and between benthic, littoral, and limnetic systems (Carpenter and Lodge 1986; Lodge et al. this volume). Many other puzzles in limnology may be solved by recognizing and quantifying the links among terrestrial, littoral, benthic, and pelagic subsystems (Lodge et al. this volume).

Complex interactions may explain the great sensitivity of community structure (in comparison with productivity and nutrient cycles) to the stress of lake acidification (Schindler et al. 1985). Community responses to other stresses such as eutrophication, introduction of exotic species, or exploitation have also produced surprises, sometimes with long time lags (Carpenter 1980; Edmondson and Litt 1982; Kitchell and Carpenter 1987; Scavia and Fahnenstiel this volume). Despite substantial changes in community structure, compensatory dynamics of functionally similar species can prevent large changes in collective processes such as primary production (O'Neill et al. 1986). On the other hand, changes in certain singularly important species lead to substantial changes in ecosystem parameters (Carpenter et al. 1987; Scavia and Fahnenstiel this volume). While community ecology advances from observations of many different taxa, the public interest is typically focused on selected economically important species. Thus, it becomes increasingly important to improve our capacity to predict responses of specific plant or animal populations.

The joint effects of nutrients and food web processes are elements of novel proposals for lake management (Shapiro 1980, Shapiro and Wright 1984) and the cascade hypothesis for temporal scaling of variance in lake productivity (Carpenter and Kitchell 1987). Some current controversies about food web dynamics in lakes and the implications of time scale are explored in many of the chapters in this volume. The recent recognition of the significance of the microbial loop adds a new dimension to the interactions of nutrient supply and food web dynamics in lakes (Porter et al. this volume; Stockner and Porter this volume; Scavia and Fahnenstiel this volume).

We emphasize the importance of lake history and the persistent or delayed effects of past events on community interactions. Equilibrium is not among the intrinsic properties of communities, but can emerge at sufficiently large spatial scales (DeAngelis and Waterhouse 1987). The small size of many lakes, within-lake heterogeneity (Lodge et al. this volume), and the constrained spatial scales imposed by hydrodynamics (Harris 1986) may limit the usefulness of equilibrium models for lake communities.

Historical effects on community interactions are clearly evident on evolutionary time scales (DeAngelis et al. 1985; Kerfoot and Lynch 1987). Numerous examples of historical effects exist on shorter time scales as well. Catastrophic disturbance of watershed vegetation produces surprisingly short-lived nutrient pulses (Bormann and Likens 1985), which nevertheless may affect lakes for long periods through internal storage and recycling processes. A strong year class of fishes regulates pelagic community structure throughout the lifetime of the fish (Carpenter et al. 1985). Introductions of exotic macrophytes (Carpenter 1980) or crayfish (Lodge and Lorman 1987) have profoundly altered littoral zone communities for periods of years to decades. The depth and duration of spring mixing set the productivity of temperate lakes for the subsequent year (Strub et al. 1985). Harris (1986) argues that temporal heterogeneity associated with hydrodynamic processes is the principal determinant of phytoplankton community structure. Temporal heterogeneity poses a major and unavoidable challenge to progress in lake ecology.

Why Lakes?

Complex interactions are very general phenomena that are important in all communities and ecosystems. Among the major types of ecological systems, lakes are especially appropriate for the study of complex interactions. Workshop participants identified eight major advantages of lakes for the study of complex interactions.

1. Lakes are bounded, and inputs and outputs are readily quantified.
2. Subsystems are readily defined and isolated for experimental purposes.
3. Certain subsystems, such as the epilimnion in summer and some benthic habitat types, are relatively homogeneous.
4. Lakes can be understood through reference to only a few dominant time scales, such as the diel cycle, annual cycle, the life cycles of certain keystone predators (cf. Carpenter this volume), and the frequencies of major watershed and/or airshed disturbances (Bormann and Likens 1985).
5. Lakes are amenable to experimentation on a variety of spatial scales, including whole-lake manipulations. The natural variability of lakes is tractable, and allows detection of major effects in large-scale experiments (Likens 1985).
6. Evolution clearly operates in an ecological context in lakes. For example, zooplankton behaviors and morphologies are affected by food web interactions through known mechanisms (Havel 1987; Kerfoot and Lynch 1987).
7. A diversity of lake ecosystem types (ranges of geologic age, size, morphometry, trophic status, thermal regime) exists over accessible spatial scales.

8. Lakes are sensors and recorders of the larger global environment. For example, paleoecological records gleaned from lake sediments reveal sweeping changes in the climate and biogeography of North America over thousands of years (Davis 1981).

Finally, we point to the great societal value of fresh water resources and fisheries. Water quality and fisheries are linked through complex interactions (Shapiro 1980; Carpenter et al. 1985) as well as major global phenomena such as acid deposition (Schindler et al. 1985). Several chapters in this book address the profound effects of fisheries management practices on plankton communities and water quality. Macrophyte management has impact on both fisheries and water quality (Carpenter and Lodge 1986). We suggest that basic research on complex interactions will lead to explanations for the variability and unpredictability that presently hamper lake management efforts.

Overview of the Book

This book reflects the evolution of our views on complex interactions before, during, and after the workshop at Notre Dame. Plenary talks presented at the workshop were designed to present examples of complex interactions as a basis for discussion. Papers derived from these plenary talks appear in three parts.

Part I, Patterns and Surprises in Lake Food Webs, includes three papers that illustrate the complicated and sometimes counterintuitive joint effects of nutrient supply and predation in lake communities. Mills and Forney show how the seasonal succession of the Oneida Lake plankton depends on the recruitment of perch each spring. Neill recounts enclosure experiments which have shown surprising effects of nutrient enrichment on food web interactions. Persson et al. provide a historical perspective on the schism between limnology and fisheries ecology, and offer a new synthesis of trophic state and food web structure. The "bottom–up" vs. "top–down" debate that has emerged from such research (Crowder et al., this volume) applies a rich data base and well-established mechanistic models to an issue that parallels the recent controversy about the importance of competition (Salt 1983).

Part 2, Microbial Links to the Classic Food Web, summarizes recent discoveries about the ecology of microbes and protozoa that may completely alter our concepts of aquatic food webs. Stockner and Porter discuss implications of microbial interactions for trophic linkage patterns and carbon cycling. Scavia and Fahnenstiel describe effects of massive changes in the fishery of Lake Michigan extending through the zooplankton and net phytoplankton to microbial and detrital processes. These papers, and the discussion group report by Porter et al., lead to questions beyond that of the role of microbes as a source or a sink for carbon. For example, what are the combinations of trophic state and food web structure that permit the microbial link to be a source or a sink?

Part 3, Multiple Causality and Temporal Pattern in Lake Systems, addresses the effects of time scale on control of system processes when multiple causal pathways exist. Bartell et al. show that both "bottom–up" and "top–down" factors control the phytoplankton, and that at certain times of the year one factor or the other appears dominant. Carpenter shows that "top–down" effects produce the greatest variance in phytoplankton at time scales near the life span of the dominant predator. These papers caution against arguments for univariate control of system processes based on studies at a single time scale, and suggest that dominant processes in lake communities may operate at time scales substantially longer than the traditional triennial granting cycle.

The remainder of the book consists of reports of discussions held at the Workshop on Complex Interactions in Lake Communities. These discussions were held in two phases. The first phase was topically oriented and conducted by chairperson–rapporteur teams who had the opportunity to plan the discussions in advance. The second phase, synthesis discussions, was conducted ad hoc and combined individuals from different topic-oriented discussions. Background information on each set of discussions introduces Parts 4 and 5, Reports from Group Discussions and Synthesis, respectively.

Acknowledgements. We thank John Barko, Tom Frost, Chris Luecke, and Robert McIntosh for comments on an earlier draft of this chapter.

References

Borgmann, U. 1987. Models on the shape of, and biomass flow up, the biomass size spectrum. Can. J. Fish. Aquat. Sci. In press.

Bormann, F. H. and G. E. Likens. 1985. Air and watershed management and the aquatic ecosystem. *In:* An ecosystem approach to aquatic ecology, ed. G. E. Likens, 436–444. New York: Springer-Verlag.

Carpenter, S. R. 1980. The decline of *Myriophyllum spicatum* in a eutrophic Wisconsin lake. Can. J. Bot. 58:527–535.

Carpenter, S. R. and J. F. Kitchell. 1987. The temporal scale of variance in limnetic primary productivity. Am. Nat. 129:417–433.

Carpenter, S. R., J. F. Kitchell, and J. R. Hodgson. 1985. Cascading trophic interactions and lake productivity. BioScience 35:634–639.

Carpenter, S. R., J. F. Kitchell, J. R. Hodgson, P. A. Cochran, J. J. Elser, M. M. Elser, D. M. Lodge, D. Kretchmer, X. He, and C. N. von Ende. 1987. Regulation of lake primary productivity by food web structure. Ecology 68:1863–1876.

Carpenter, S. R. and D. M. Lodge. 1986. Effects of submersed macrophytes on ecosystem processes. Aquat. Bot. 26:341–370.

DeAngelis, D., J. A. Kitchell, and W. M. Post. 1985. The influence of naticid predation on evolutionary strategies of bivalve prey: conclusions from a model. Am. Nat. 126:817–842.

DeAngelis, D. L. and J. C. Waterhouse. 1987. Equilibrium and nonequilibrium concepts in ecological models. Ecol. Monogr. 57:1–21.

Davis, M. B. 1981. Quaternary history and the stability of forest communities. *In*: Forest Succession, ed. D. C. West, H. H. Shugart, and D. B. Botkin. New York: Springer Verlag.

Diamond, J. and T. Case. 1986. Ecological Communities. New York: Harper and Row.

Edmondson, W. T. 1979. Lake Washington and the predictability of limnological events. Arch. Hydrobiol. Beih. 13:234–241.

Edmondson, W. T. and A. H. Litt. 1982. *Daphnia* in Lake Washington. Limnol. Oceanogr. 27:272–293.

Havel, J. E. 1987. Predator-induced defenses: A review. *In*: Predation: Direct and indirect impacts on aquatic communities, ed. W. C. Kerfoot and A. Sih, 263–278. Hanover: University Press of New England.

Kerfoot, W. C. and M. Lynch. 1987. Branchiopod communities: Associations with planktivorous fish in space and time. *In*: Predation: Direct and indirect impacts on aquatic communities, ed. W. C. Kerfoot and A. Sih, 367–378. Hanover: University Press of New England.

Kerfoot, W. C. and A. Sih, eds. 1987. Predation: direct and indirect impacts on aquatic communities. Hanover: University Press of New England.

Kitchell, J. F. and S. R. Carpenter. 1987. Piscivores, planktivores, fossils, and phorbins. *In*: Predation: Direct and indirect impacts on aquatic communities, ed. W. C. Kerfoot and A. Sih, 132–146. Hanover: University Press of New England.

Likens, G. E. 1985. An experimental approach for the study of ecosystems. J. Ecol. 73:381–396.

Lodge, D. M. and J. G. Lorman. 1987. Reductions in submersed macrophyte biomass and species richness by the crayfish *Orconectes rusticus*. Can. J. Fish. Aquat. Sci 44:591–597.

O'Neill, R. V., D. L. DeAngelis, J. B. Waide, and T. F. H. Allen. 1986. A hierarchical concept of ecosystems. Princeton: Princeton University Press.

Salt, G. W., ed. 1983. A round table on research in ecology and evolutionary biology. Am. Nat. 122:583–705.

Schindler, D. W., K. H. Mills, D. F. Malley, D. L. Findlay, J. A. Shearer, I. J. Davies, M. A. Turner, G. A. Linsey, and D. R. Cruikshank. 1985. Long-term ecosystem stress: The effects of years of experimental acidification on a small lake. Science 228:1395–1401.

Shapiro, J. 1980. The importance of trophic-level interactions to the abundance and species composition of algae in lakes. *In*: Hypertrophic Ecosystems, ed. J. Barica and L. Mur, 105–115. The Hague: Dr. W. Junk Publishers.

Shapiro, J. and D. I. Wright. 1984. Lake restoration by biomanipulation. Freshwater Biol. 14:371–383.

Sheldon, R. W., W. H. Sutcliffe, and M. A. Paranjape. 1977. Structure of pelagic food chains and the relationship between plankton and fish predation. J. Fish. Res. Bd. Can. 34:2344–2353.

Sprules, W. G. and M. Munawar. 1986. Plankton size spectra in relation to eco-system productivity, size, and perturbation. Can. J. Fish. Aquat. Sci. 43:1789–1794.

Strub, P. T., T. Powell, and C. R. Goldman. 1985. Climatic forcing: Effects of El Niño on a small, temperate lake. Science 227:55–57.

Tilman, D., S. S. Kilham, and P. Kilham. 1982. Phytoplankton community ecology: The role of limiting nutrients. Ann. Rev. Ecol. Syst. 13:349–372.

Werner, E. E. and J. F. Gilliam. 1984. The ontogenetic niche and species inter-actions in size-structured populations. Ann. Rev. Ecol. Syst. 15:393–425.

Wetzel, R. G. 1979. The role of the littoral zone and detritus in lake metabolism. Arch. Hydrobiol. Beih. 13:145–161.

Wetzel, R. G. and H. L. Allen. 1970. Functions and interactions of dissolved organic matter and the littoral zone in lake metabolism and eutrophication. *In*: Productivity problems of freshwaters, ed. Z. Kajak and A. Hillbricht-Ilkowska, 333–347. Warsaw: PWN Polish Scientific Publishers.

Wetzel, R. G., P. H. Rich, M. C. Miller, and H. L. Allen. 1972. Metabolism of dissolved and particulate detrital carbon in a temperate hard-water lake. Mem. Ist. Ital. Idrobiol. 29 (suppl.):185–243.

Part 1
Patterns and Surprises in Lake Food Webs

CHAPTER 2

Trophic Dynamics and Development of Freshwater Pelagic Food Webs

Edward L. Mills and John L. Forney

Introduction

The open water or pelagic zones of lakes are characterized by complex interactions that sometimes include edge effects from nearshore and near-bottom areas (Hutchinson 1957). The freshwater pelagia occur in water bodies ranging from small lakes to the world's great lakes and communities inhabiting this zone are affected by small- to large-scale physical, chemical, and biological processes. Community interactions in pelagic food webs are trophically dynamic and are governed by nutrient limitation, competition, predation, and other ecological forces. Rate processes in pelagic food webs are generally faster than those in terrestrial systems (Paine 1980) and consequently producer controlled and consumer controlled trophic level interactions are more readily observed. Freshwater pelagic food webs have been studied intensively in this century and there is growing awareness of their complexity and uniqueness in aquatic ecology (Persson et al. this volume).

In pelagic ecosystems, most primary production occurs at microscopic levels and most predators are larger than their prey. The biological structure of freshwater pelagic ecosystems is, however, complex, and is regulated hierarchically through both biotic and abiotic mechanisms (Carpenter et al. 1985). Potential productivity at all trophic levels is set by nutrient supply, while biological structure within the food web determines how production is packaged. Whether the biomass and abundance of organisms in pelagic food webs is controlled from below by producers or from above by consumers is now subject to debate. In this paper, we summarize empirical data for pelagic ecosystems supporting the producer and consumer models. We also examine the pelagic food web of Oneida Lake, New York and illustrate community dynamics and both seasonal and long-term development of this food web. Finally, we discuss research areas that require further development in order to better understand complex interactions in aquatic food webs.

Consumer vs. Producer Control in Pelagic Food Webs

Community structure, biomass, and production in pelagic food webs are influenced by both producers and consumers. While energy can only flow to higher trophic levels, consumers can influence the flow rate and how the energy is used. Currently, there are two opposing views of how trophic level biomass in pelagic food webs is controlled: from below by producers, or from above by consumers.

Producer control has evolved as a logical extension of the trophic state model (Hutchinson 1973; Wetzel 1983; Persson et al. this volume) and gained added acceptance through studies of eutrophication. Empirically derived relationships supporting the producer controlled model include regressions of (1) chlorophyll concentration or Secchi disc transparency on the independent variable phosphorus (Sakamato 1966; Dillon and Rigler 1974; Jones and Bachman 1976; Oglesby and Schaffner 1978; Janus and Vollenweider 1981; Hanson and Peters 1984; Pace 1984); (2) fish yield on the independent variables total phosphorus (Hanson and Leggett, 1982), phytoplankton standing stock or production (Melack 1976; Oglesby 1977; Jones and Hoyer 1981; Liang et al. 1981); (3) mean zooplankton biomass on the independent variables phytoplankton biomass or chlorophyll a (McCauley and Kalff 1981; Mills and Schiavone 1982; Hanson and Peters 1984; Canfield and Watkins 1984); (4) an index of planktivore biomass or zooplanktivore biomass on the independent variable zooplankton biomass (Mills and Schiavone 1982); and (5) an index of piscivore biomass on an index of planktivore biomass (Mills and Schiavone 1982). Each of these relationships suggest that the biomass of the upper trophic level is dependent on biomass at a lower trophic level.

The consumer controlled model has generated the theories of biomanipulation (Shapiro et al. 1982; Shapiro and Wright 1984) and cascading trophic interactions (Carpenter et al. 1985). While potential biomass is dependent on mixing, nutrient supply, and other abiotic factors, actual biomass and biological structure are controlled from above by consumers. Over the last decade, numerous studies have supported the concept that the biomass at each trophic level can be controlled from above by consumers. Results of these studies have shown that (1) a rise in piscivores reduces planktivore abundance (Bonar 1977; Holcik 1977); (2) an increase in planktivore biomass impacts zooplankton species composition and zooplankton biomass (Hrbacek 1962; Brooks and Dodson 1965; Galbraith 1967; Hall et al. 1970; Hutchinson 1971; Stenson 1972, 1976; O'Brien and DeNoyelles 1974; Anderson et al. 1978; Hurlbert and Mulla 1981; Mills and Forney 1983; McQueen and Post 1985; Vijverberg and van Densen 1984; McQueen et al. 1986); and (3) an increase in zooplankton biomass can reduce phytoplankton biomass (Porter 1977; Lynch 1979; Lynch and Shapiro 1981; Shapiro et al. 1982; McCauley and Kalff 1981; Shapiro and

Wright 1984), but the relationship between zooplankton biomass and primary production can be nonlinear (Bergquist and Carpenter 1986).

Recently, McQueen et al. (1986) have shown that maximum biomass at each trophic level in freshwater pelagic ecosystems is controlled from below by nutrient availability. However, at the top of the food chain, predator mediated interactions are strong but weaken with each trophic level down the food chain. Similarly, Mills et al. (1987) have found that fish predation impacts zooplankton and phytoplankton but that the response becomes less predictable with trophic distance between variables.

In pelagic food webs, producers and consumers are functioning simultaneously and therefore the question remains, which force dominates the biomass and size structure of freshwater pelagic food webs? McQueen et al. (1986) contend that maximum attainable biomass is set by lake productivity, and that consumer forces are stronger in eutrophic lakes while producer forces are stronger in oligotrophic lakes. In oligotrophic systems, consumer effects are strongest at the primary producer and herbivore levels, but with increasing eutrophy, consumer forces move up the food chain to piscivores. However, the productivity threshold where producer controlled forces on biomass and size structure give way to consumer control remains undefined.

Food Web Dynamics and Development: A Case History of Oneida Lake, New York

The System

Oneida Lake is a shallow eutrophic lake with a surface area of 20,700 ha, located on the Ontario Lake Plain of central New York. Mean depth is 6.8 m and 26% of the area is less than 4.3 m in depth. The lake is alkaline and water chemistry is strongly influenced by nutrient rich streams that flow from the south over outcroppings of Onondaga limestone and through the fertile and highly populated Ontario Lake Plain (Mills et al. 1978).

Nearly 60 species of fish inhabit Oneida Lake, with walleye (*Stizostedion vitreum vitreum*) being the dominant predator and gamefish. Studies began in 1957 to estimate population dynamics of walleye and their impact on prey populations (Forney 1980). Since the 1970s, limnological studies have been coupled with fishery studies to measure how species level interactions are propagated through the food chain.

The Oneida Lake Food Chain

The walleye is the dominant piscivorous fish and it shares the limnetic region with yellow perch (*Perca flavescens*) and smaller numbers of white perch (*Morone americana*). These three species comprise over 90% of the catch in variable mesh gillnets (Clady 1978). Clupeids and large bodied obligate planktivores are scarce and in their absence, age-0 yellow perch along with age-0 white perch and young-of-the-year of other species serve as the primary link in the transfer of energy from secondary to higher trophic levels. Because cohorts of age-0 fish vary in initial size and are subject to different rates of predation, biomass of these planktivores fluctuates both seasonally and annually.

Zooplankton play a pivotal role, serving as food for planktivorous fish and consumers of phytoplankton. Nineteen zooplankton species have been identified, and their relative abundance varies both seasonally and annually. Typically, daphnids dominate in the spring, and whether this community persists through the summer depends mostly on the abundance of young yellow perch. In years when young fish are abundant, *Chydorus*, *Bosmina*, *Diaphanosoma*, and *Ceriodaphnia* often replace the larger bodied daphnids. *Diaptomus*, *Cyclops*, and *Epischura* and other copepods coexist with *Daphnia* and generally pulse during the spring and early summer.

Temporal succession of the phytoplankton community is well defined, with diatoms dominating soon after ice-out and yielding to a late spring assemblage of small flagellates. Cyanophytes usually become abundant during the summer, sometimes concurrent or alternating with blooms of filamentous diatoms, producing periods of maximum phytoplankton biomass. These blooms may persist into the fall, after which the diversity of algal species usually increases and the biomass declines.

Annual Development of the Food Chain

Development of the Oneida Lake food chain is an annual event, and the pattern of maturation is governed by the abundance of age-0 yellow perch. Here, we trace the seasonal development of a yellow perch cohort and examine the impact of differences in cohort biomass on higher and lower trophic levels.

Yellow perch spawn nearshore in late April. Eggs hatch in early May and larvae concentrate in open water at depths of 0–5 m. Young reach 18 mm by mid- to late June, and transition from the pelagic to demersal mode occurs in July at a length of 30–40 mm. By early August, young weigh 1–2 g (Forney 1980). Number of eggs spawned has varied within a relatively narrow threefold range between 1961 and 1986 (Table 2.1). However, climate induced variation in early survival (Clady 1976) coupled with predator mediated depensatory mortality in late summer (Forney 1971) leads to rapid divergence in year–class abundance. By the time yellow perch cohorts in 1961–86 attained 18 mm there were 24-fold differences, and differences in year–class size reached 100-fold by fall. These annual fluctuations in cohort abundance of the primary planktivore lead to complex interactions that are transmitted through the food chain.

TABLE 2.1. Abundance of yellow perch (number/ha) estimated at successive stages of cohort development, spring through fall 1961–1986

Year class	Eggs (millions)	Age-0 yellow perch		
		18 mm[1]	Aug 1[2]	Oct 15[2]
1961	2.07	– –	28,600	2,850
1962	2.24	– –	21,100	4,350
1963	2.58	– –	18,500	780
1964	2.60	– –	26,000	4,990
1965	3.00	140,000	21,700	2,610
1966	2.68	40,000	12,000	150
1967	2.93	61,000	24,500	2,100
1968	3.02	– –	36,000	6,700
1969	2.84	68,000	7,200	210
1970	2.83	69,000	14,000	900
1971	2.16	217,000	37,700	3,520
1972	2.63	110,000	2,500	100
1973	3.26	18,000	1,200	510
1974	3.28	30,000	3,100	320
1975	2.46	152,000	21,400	450
1976	2.28	37,000	6,500	190
1977	1.96	65,000	14,700	4,220
1978	2.16	– –	6,100	180
1979	1.61	103,000	13,400	360
1980	4.83	132,000	8,900	500
1981	4.44	208,000	21,200	2,590
1982	2.89	353,000	16,100	934
1983	2.33	46,000	9,600	664
1984	2.78	16,000	5,030	896
1985	– –	91,000	8,360	2,720
1986	– –	14,637	509	66
Mean	2.74	98,532	14,842	1,687
Coefficient of variation	0.26	0.85	0.69	1.08

[1]Population estimate adjusted to 18 mm in 1965–77; point estimates 1978–86.
[2]Estimated from c/f in trawls.

Walleye predation on young yellow perch begins when cohorts reach a length of 18 mm, and in most years few young survive to the fall (Table 2.1). Biomass of yellow perch cohorts increases through June, but in late summer consumption by walleye exceeds production and yellow perch biomass declines. Alternate prey are usually scarce, and somatic production by age-3 and older walleye is roughly proportional to the production attained by each cohort of young yellow perch (Mills et al. 1987). High production of yellow perch can enhance walleye recruitment by reducing intensity of cannibalism (Chevalier 1973).

Impact of yellow perch cohorts on lower trophic levels is best illustrated by contrasting events in a year of high and low yellow perch abundance (Figs. 2.1 and 2.2). In 1975, age-0 yellow perch biomass reached 32 kg/ha in July. *Daphnia pulex* pulsed in May and then collapsed under intense yellow perch predation (Mills and Forney 1983). Disappearance of *D. pulex* caused a dramatic decline in mean size of the crustacean zooplankton community and a shift to a midsummer mixture of copepods and smaller cladocerans. The switch from large bodied to small bodied grazers was followed in August by a peak in net phytoplankton, but biomass of nannoplankton (< 50 microns) was unchanged. In contrast, yellow perch biomass in 1976 never exceeded 12 kg/ha and the *Daphnia pulex* population remained high throughout the summer. Species composition of the zooplankton community showed little change and mean size of the crustacean zooplankton community remained relatively constant during the summer. Presence of *Daphnia pulex* failed to reduce biomass of net phytoplankton in 1976 but nannobiomass was greatly reduced. However, changes in body size and species composition of the zooplankton community between 1975–1976 had little effect on mean total biomass of phytoplankton. Mean biomass of zooplankton in June through September averaged 141.4 µg/l in 1975 and 220.4 µg/l in 1976. Despite differences in zooplankton biomass, total phytoplankton biomass remained relatively constant and averaged 4249 µg/l and 4057 µg/l June through September of 1975 and 1976, respectively. However, nannoplankton declined from nearly 1000 µg/l to 185 µg/l in late summer of 1976, but a similar decline did not occur in 1975.

Trophic level interactions observed in 1975–1976 were further substantiated through examination of a larger data set. For the period 1975–1986, significant ($P < 0.05$) negative correlations were found between total zooplankton biomass and yellow perch biomass ($r^2 = 0.41$; d.f. $= 9$), and between nannoplankton biomass and total zooplankton biomass ($r^2 = 0.54$; d.f. $= 9$), but no relationship was observed between total net phytoplankton biomass and total zooplankton biomass. Apparently, zooplankton are unable to effectively exploit the abundant supply of net phytoplankton in late summer, which consists of blue-greens and other large phytoplankters. This mismatch in size of producer and consumer alters trophic level biomass and creates a bottleneck which reduces energy transfer through the food chain (cf. Neill this volume, Stein et al. this volume).

The key feature in the seasonal development of the walleye–yellow perch–zooplankton food chain is the consistent collapse of the *Daphnia pulex* population in years when biomass of age-0 yellow perch exceeds 20 kg/ha (Mills et al. 1987). Apparently, this threshold is species specific and not broadly applicable to other young planktivores in Oneida Lake. *Daphnia pulex* remained abundant in late summer of 1980 and 1983, although biomass of age-0 white perch approached or exceeded 20 kg/ha (Fig. 2.3). Abundance of white perch was estimated from the catch in

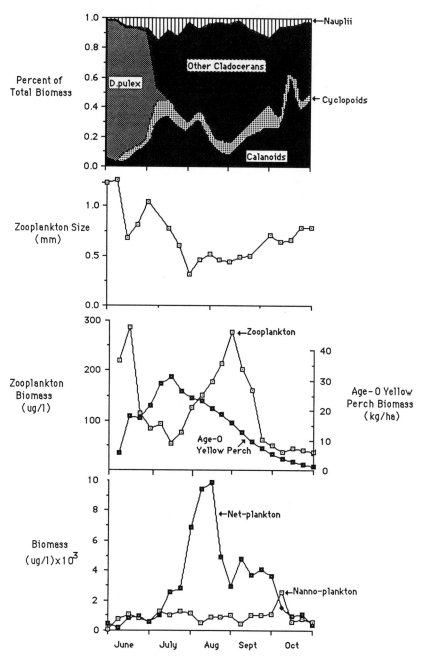

FIGURE 2.1. Comparison of seasonal changes in biomass of phytoplankton, zoo-plankton, and age-0 yellow perch, mean length of zooplankton, and proportion of total biomass represented by zooplankton groups in Oneida Lake during 1975.

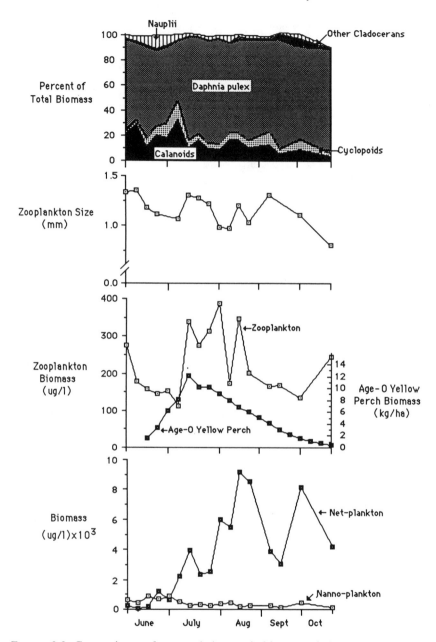

FIGURE 2.2. Comparisons of seasonal changes in biomass of phytoplankton, zoo-plankton, and age-0 yellow perch, mean length of zooplankton, and proportion of total biomass represented by zooplankton groups in Oneida Lake during 1976.

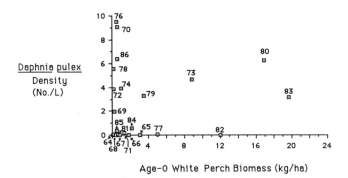

FIGURE 2.3. Biomass of age-0 white perch determined from trawl catches, averaged August–October, compared with density of *Daphnia pulex*, averaged August–September, in Oneida Lake, 1964–1986.

trawls fished during the day when their vulnerability to capture is relatively low (Forney 1974). Had samples been taken at night, estimates of biomass in 1980 and 1983 would probably have exceeded 50 kg/ha. Evidently, age-0 white perch do not have the same impact on *D. pulex* as do young yellow perch. Compared to yellow perch, white perch spawn later, attain their maximum biomass in September, and achieve a higher weight and length by fall. These differences in ontogeny coupled with differences in spatial distribution and prey selection may make *D. pulex* less susceptible to or more tolerant of predation by white perch than yellow perch.

Stability of Food Webs

The fish community and its impact on structure and function of the Oneida Lake ecosystem has been the subject of research for nearly 30 yr. Here, we trace the dynamics of the fish and daphnid community over this period and speculate on the role of predation in its development.

Walleye and yellow perch comprise a demographically simple association, but the advent of white perch and gizzard shad in Oneida Lake may pose a threat to the stability of this association. Gizzard shad were first reported from Oneida Lake in 1954 when vast schools of young appeared along shore and attracted public attention. The irruption was brief and there was no further evidence of successful reproduction until 1973, when a few young were taken in trawls and seines. In subsequent years, numbers of adult caught in gillnets increased (Table 2.2) and large schools of age-0 gizzard shad reappeared in 1984–1986. White perch invaded the lake from the Mohawk–Hudson River system in the late 1940s (Scott and Christie 1963) and produced several strong year-classes in the 1950s. Annual catches of yearling and older white perch in gillnets exceeded 200 in the early 1960s, then declined to less than 100 in the early 1970s but increased to over 500 in the late 1970s. Catches of age-0 white perch in

TABLE 2.2. Total catch of yellow perch, white perch and gizzard shad in 15 gillnet sets from July to October, 1961–1986

Year	Catch		
	Yellow perch	White perch	Gizzard shad
1961	1379	98	0
1962	1491	142	0
1963	1670	198	0
1964	1611	233	0
1965	1938	254	0
1966	1346	129	0
1967	1540	106	0
1968	1937	73	0
1969	1510	65	0
1970	1866	176	0
1971	830	45	0
1972	926	24	0
1973	1531	28	1
1974	– –	– –	– –
1975	1427	415	0
1976	1434	501	0
1977	1433	207	1
1978	1053	100	1
1979	1649	509	0
1980	2838	351	7
1981	2420	963	1
1982	1722	486	0
1983	1166	416	8
1984	1499	424	12
1985	750	666	60
1986	1074	921	25

trawls followed the same temporal pattern of abundance as adults. On the positive side, the availability of gizzard shad and white perch has increased walleye production, but irruptions of these fish could destabilize the fish community. The only historical example we have to draw upon comes from the mid-1950s. Wide oscillations in abundance and harvest of walleye followed the gizzard shad irruption in 1954, and dominant year–classes of white perch, white bass, and yellow perch produced in 1954 persisted into the 1960s (Forney 1980).

Structural changes in the fish community were matched by shifts in species composition of the zooplankton community. *Daphnia galeata* and *D. retrocurva* were the most abundant species from 1964–1969 (Fig. 2.4). *Daphnia pulex* dominated throughout the 1970s, but *D. galeata* reemerged from relative obscurity to become the predominant daphnid after 1981. Disappearance and reappearance of *Daphnia galeata* was associated with

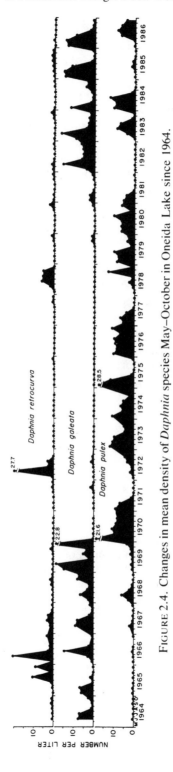

FIGURE 2.4. Changes in mean density of *Daphnia* species May–October in Oneida Lake since 1964.

changes in abundance of age-0 yellow perch and other forage fish (Table 2.3). Forage-size fish included all age-0 percids, serranids, clupeids, and centrarchids and older trout perch (*Percopsis omiscomaycus*), tessellated darter (*Etheostoma olmstedi*), and log perch (*Percina caprodes*). Dominance of *Daphnia galeata* when age-0 yellow perch biomass was high and appearance of the larger bodied *D. pulex* in 1970–1980 when yellow perch and all age-0 forage sized fish biomass was low implicates predation as the factor regulating species composition of the daphnid community. But, the appearance of *D. galeata* in the 1980s despite continued low abundance of age-0 yellow perch and only a modest increase in the catch of other forage size fish is more difficult to explain.

Age-0 yellow perch contributed 90% of the total forage fish biomass in 1964–1969, 66% in 1970–1980, and only 47% in 1981–1986. The decline in yellow perch was accompanied by a substantial increase in other small planktivores (primarily white perch and gizzard shad), although their increase was not reflected in estimates of forage fish biomass because most species were less vulnerable to capture in trawls than yellow perch (Forney 1974). Differences in vulnerability were most striking in 1984 and 1985, when age-0 gizzard shad replaced yellow perch as the primary item in the diet of walleye, but their contribution to the trawl catch was negligible. Even though estimates of age-0 fish abundance are conservative, it is likely that total biomass of planktivorous age-0 fish since 1980 has approached that prior to 1970. While high biomass of white perch in 1980 and 1983 failed to suppress *Daphnia pulex*, predation may have prevented *D. pulex* from monopolizing algal resources and excluding the congeneric *D. galeata* (Kerfoot 1987).

The simple walleye–yellow perch association of the 1960s has become more complex as successful reproduction by subordinate species has increased planktivore diversity. Large year–classes of age-0 white perch and gizzard shad have increased the potential for interspecific competition,

TABLE 2.3. Mean density of *Daphnia pulex* and *Daphnia galeata* averaged August through September, mean biomass of age-0 yellow perch averaged August through October, and all forage size age-0 fish averaged July through October in Oneida Lake between 1964 and 1986

	Mean density		Mean biomass	
	Daphnia pulex	*Daphnia galeata*	Age-0 yellow perch	All age-0 fish
Period	(#/L)		(kg/ha)	
1964–1969	0.34	3.38	19.0	20.7
1970–1980	4.21	0.12	8.0	12.1
1981–1986	1.76	2.57	7.5	15.9

increased complexity of interactions, and possibly decreased stability. Recent changes in the daphnid community may reflect destabilizing effects of a changing fish community. Since 1981, *D. pulex* had been co-dominant with *D. galeata* in 4 of 6 yr, maximum *D. galeata* density has equaled or exceeded *D. pulex* density in all years, and density of *D. galeata* has peaked before *D. pulex* in 4 of 6 years. There were signs that both daphnid species coexisted in only 5 of 17 years prior to 1981. In all cases, the subdominant daphnid species never exceeded a density of 5 per liter, and the density of *D. galeata* exceeded that of *D. pulex* in 6 of 17 years. Continued variability in the daphnid community combined with a more diverse forage fish community could have adverse effects on the yellow perch population and its linkage as a primary forage for the walleye.

Consumer–Producer Coupling

In general, the consumer model predicts an inverse relationship between producer and consumer (as independent variable) while the producer model predicts a direct linkage between producer and consumer (as dependent variable). To quantify interactions in the Oneida Lake food web and assess their strength according to the producer and consumer models, linear regressions were fitted to annual estimates of abundance at each trophic level (Table 2.4). At the top of the food web, walleye control yellow perch abundance by consuming most members of each year's cohort (Forney 1977, Neilsen 1980). An inverse relation between mortality of age-0 yellow perch and walleye biomass implies that an increase in walleye abundance leads to more rapid depletion of yellow perch cohorts (Mills et al. 1987). However, biomass of age-4 and older walleye in early summer is proportional to production of age-0 yellow perch in earlier years, providing evidence that predator biomass is regulated from below by forage fish abundance (Table 2.4). Thus, the intensity of consumer control exerted by walleye in a given year reflects the cumulative influence of yellow perch abundance in earlier years on walleye recruitment (Chevalier 1973), growth (Forney 1977), and rates of walleye harvest (Forney 1980).

Yellow perch cohorts which are numerically abundant in early summer usually attain the highest first year production (Forney 1971), and predation by these cohorts often depletes the daphnid population by midsummer (Mills and Forney 1983). The inverse relationship between age-0 yellow perch production (independent variable) and crustacean zooplankton size is evidence for consumer regulation of zooplankton composition (Table 2.4). However, the direct relationship between biomass of *Daphnia pulex* (independent variable) averaged August–September and the length young yellow perch attained by fall suggests a bottom–up effect of producers on consumers (Table 2.4). Consequently, competition by yellow perch for daphnids creates a feedback loop which not only influences yellow perch

TABLE 2.4. Predictive equations between individual trophic levels illustrating top–down consumer effects and producer induced feedback loops in Oneida Lake, New York

Independent × dependent variable (Consumer × producer)		Years (No.)	Relationship	r^2	Significance level (p)
		Top–down predator control			
Jul–Oct Age-0 yellow perch production (g/m2)	×	Jul–Oct Mean crustacean zooplankton size (mm)	11	$Y = 1.069 - 0.125 (x)$	0.58	< 0.001
Aug–Sept *Daphnia pulex* biomass (μg/l)	×	Nannophyto-plankton biomass (μg/l)	10	$Y = 803.41 - 3.09 (x)$	0.81	< 0.001
(Producer × consumer)					
		Producer induced feedback loops			
Age-0 yellow perch production averaged 5 yr preceding April 1 (g/m2)	×	Walleye biomass April 1 (kg/ha)	19	$Y = 7.51 + 2.70 (x)$	0.33	< 0.05
Aug–Sept *Daphnia pulex* biomass (μg/l)	×	Oct. 15 weight Age-0 yellow perch (g)	12	$Y = 3.54 + 0.01 (x)$	0.38	< 0.05
May–Jun nannophyto-plankton biomass (μg/l)	×	May–Jun *Daphnia pulex* brood size	8	$Y = 2.625 + 0.002 (x)$	0.27	< 0.02

production but also impacts their recruitment by prolonging the period of their vulnerability to walleye predation (Neilsen 1980).

At the base of the food web, biomass of *Daphnia* is inversely proportional to nannoplankton biomass; nearly 80% of the variance in nannoplankton biomass can be attributed to changes in biomass of *Daphnia pulex* (Table 2.4). On the other hand, producer forces influence *Daphnia* recruitment as evidenced by the direct influence of nannoplankton biomass on brood size of *Daphnia pulex* in the spring (Table 2.4). However, a significant biomass of phytoplankton (primarily blue-greens) is unavailable to *Daphnia* during the summer and fall, creating a bottleneck in the food web and a potential loss of *Daphnia* production (Wagner 1983; Neill this volume; Stein et al. this volume). The decrease in *Daphnia* production impedes the flow of energy to yellow perch and results in submaximal production of walleye.

In Oneida Lake, walleye prey on young and yearling yellow perch, and the impact is transmitted through several trophic levels (Mills et al. 1987). However, it would appear that in this lake both consumer and producer forces operate in concert. The consumer force controls biological structure, while the producer force regulates production at each trophic level. The question remains, which of these forces dominates in Oneida Lake? The only evidence that predator mediated impacts are stronger than producer induced feedback loops is associated with both the percentage of the variance explained and the significance of interactions at each trophic level (Table 2.4). For instance, top–down predator control accounted for 58–81% of the variance ($P < 0.001$) in consumer–producer relations where the consumer was the independent variable, compared to 27–38% in producer–consumer relationships ($0.02 < P < 0.05$). These results are consistent with those that support consumer control in pelagic systems (McQueen et al. 1986). However, even though there is a higher proportion of unexplained variability in bottom-up relationships, these forces influence the recruitment and production processes that regulate the flow of energy to piscivores.

In summary, walleye, yellow perch, and daphnids in Oneida Lake have coexisted for many decades, and during this period they have had time to adapt to certain short- and long-term perturbations. Here, consumer–producer forces are closely coupled and strong feedback loops are evident. In effect, consumer forces structure the food web while competition for food resources creates feedback loops which regulate trophic level production. The strong top–down forces frequently observed in ecosystems altered by fish manipulations may reflect the short time scale of most studies (Carpenter et al. 1987). Over a longer period, producer-driven feedback loops may develop as species interactions lead to a reorganization of the ecosystem, but this possibility remains an open question.

Ecological Implications

Walleye predation suppresses prey species, but climatic and other short-term disturbances cause wide annual fluctuations in cohort strength of yellow perch and other age-0 planktivores. The relatively long life span and resultant age structure of fish populations reduces the destabilizing effects of changes in prey abundance on predator stocks, and contributes to a tenuous equilibrium between walleye and yellow perch which has persisted for over 30 yr. Despite high annual variability in both predator recruitment and forage abundance, our findings indicate that both consumer and producer forces function in concert to structure and regulate biomass. Predation appears to control community structure while competition for available food resources acts as a self-regulating mechanism controlling production at each trophic level (Fig. 2.5). In manipulated systems where excess predators are added, trophic level changes are manifested through top–down forces. In fact, trophic level responses resulting from short-term piscivore introductions have been shown to have almost

PREDATOR CONTROL

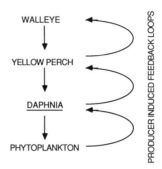

FIGURE 2.5. Illustration of top–down predator control and bottom–up consumer control feedback loops in Oneida Lake.

immediate impact on algal populations and water clarity (Shapiro and Wright 1984; McQueen et al. 1986; Carpenter et al. 1985, 1987). On the other hand, experience from Oneida Lake suggests that as predator and prey approach balance, trophic level interactions become self-regulating through both producer and consumer controlled forces. Such self-regulating mechanisms, however, can be detected only through long-term study; consequently, results from short-term ecological manipulations represent only the initial phase in the development of trophic relationships in freshwater pelagic ecosystems.

Concepts of how pelagic systems function have changed dramatically in recent decades. Currently, views of trophic interactions and their relations to ecosystem structure and function are more holistic (Persson et al. this volume). Research approaches to achieve ecosystem understanding include experimental pond manipulations (Werner and Hall 1974; Werner et al. 1983), large-scale lake manipulations (Shapiro and Wright 1984; Carpenter et al. 1987), small-scale within-lake manipulations (McQueen et al. 1986), long-term whole-lake records (Scavia et al. 1986; Mills et al. 1987), and examination of the paleolimnological record (Kitchell and Kitchell 1980; Kitchell and Carpenter 1987). However, each of these approaches toward ecosystem understanding involves different time and space scales, and differences in scale influence interpetation of trophic level interactions (Frost et al. this volume). Unfortunately, most of our knowledge of trophic dynamics in aquatic ecosystems is based on short-term manipulations which may have limited applicability to more mature ecosystems. Consequently, several areas of research require further attention over the next few decades. High variability in predator recruitment and prey abundance in aquatic systems has obscured the biotic mechanisms which contribute to community stability. Relationships among recruitment, predation, and competition must, therefore, be examined jointly to establish which force or forces are controlling ecosystem structure and function. In eutrophic systems such as Oneida Lake, information is needed on how bottlenecks in the flow of energy created by grazer resistant algae are circumvented. Obviously, species interactions (including predator–prey and predator–predator) and particle size are important when considering how food webs develop. Therefore, concerns of how species diversity, production, and trophic level efficiency affect community stability and resiliency under stress require further attention. The stability of ecosystems altered by predator introductions remains unknown; therefore, threshold levels necessary to maintain community stability need to be established. In order to foster an understanding of ecological interactions and biological coupling in aquatic systems, two requirements will be necessary: a commitment to long-term research and an interdisciplinary approach to science.

Acknowledgements. This research was supported by research grants from Cornell University and the New York State Department of Environmental Conservation. Contribution No. 102 of the Cornell Biological Field Station.

References

Anderson, G. H., H. Berggren, G. Gronberg, and C. I. Gelin. 1978. Effects of planktivorous fish on organisms and water chemistry in eutrophic lakes. Hydrobiologia 59:9–15.

Bergquist, A. M. and S. R. Carpenter. 1986. Limnetic herbivory: Effects on phytoplankton populations and primary production. Ecology 67:1351–1360.

Bonar, A. 1977. Relations between exploitation, yield, and community structure in Polish pikeperch (*Stizostedion lucioperca*) lakes, 1966–71. J. Fish. Res. Board Can. 34:1576–1580.

Brooks, J. L. and S. I. Dodson. 1965. Predation, body size and the composition of plankton. Science (Washington, D.C.) 150:28–35.

Canfield, D. E. and C. E. Watkins. 1984. Relationships between zooplankton abundance and chlorophyll *a* concentration in Florida lakes. J. Freshwater Ecol. 2:335–344.

Carpenter, S. R., J. F. Kitchell, and J. R. Hodgson. 1985. Cascading trophic interactions and lake productivity. Bioscience 35:634–639.

Carpenter, S. R., J. F. Kitchell, J. R. Hodgson, P. A. Cochran, J. J. Elser, M. M. Elser, D. M. Lodge, D. Kretchmer, X. He, and C. N. von Ende. 1987. Regulation of lake ecosystem primary productivity by food web structure in whole-lake experiments. Ecology. In press.

Chevalier, J. R. 1973. Cannibalism as a factor in first year survival of walleyes in Oneida Lake. Trans. Am. Fish. Soc. 102:739–744.

Clady, M. D. 1976. Influence of temperature and wind on the survival of early stages of yellow perch, *Perca flavescens*. J. Fish. Res. Bd. Canada 33:1887–1893.

Clady, M. D. 1978. Structure of fish communities in lakes that contain yellow perch, *Perca flavescens*. Amer. Fish. Soc. Spec. Publ. 11:109–113.

Dillon, P. J. and F. H. Rigler. 1974. The phosphorous–chlorophyll relationship in lakes. Limnol. Oceanogr. 19:767–773.

Forney, J. L. 1971. Development of dominant year classes in a yellow perch population. Trans. Am. Fish. Soc. 100:739–749.

Forney, J. L. 1974. Interactions between yellow perch abundance, walleye predation, and survival of alternate prey in Oneida Lake, New York. Trans. Am. Fish. Soc. 103:15–24.

Forney, J. L. 1977. Evidence of inter- and intraspecific competition as factors regulating walleye (Stizostedion vitreum vitreum) biomass in Oneida Lake, New York. J. Fish. Res. Bd. Canada 34:1812–1820.

Forney, J. L. 1980. Evolution of a management strategy for walleye in Oneida Lake, New York. New York Fish and Game Journal 27:105–141.

Galbraith, M. G. 1967. Size selective predation of Daphnia by rainbow trout and yellow perch. Trans. Am. Fish. Soc. 96:1–10.

Hall, D. J., W. E. Cooper, and E. E. Werner. 1970. An experimental approach to the production dynamics and structure of freshwater animal communities. Limnol. Oceanogr. 15:839–929.

Hanson, J. M. and W. C. Leggett. 1982. Empirical prediction of fish biomass and yield. Can. J. Fish. Aquat. Sci. 39:257–263.

Hanson, J. M. and R. H. Peters. 1984. Empirical prediction of crustacean zooplankton biomass and profundal macrobenthos biomass in lakes. Can. J. Fish. Aquat. Sci. 41:439–445.

Holcik, J. 1977. Changes in fish community of Klicava reservoir with particular reference to Eurasian perch (Perca fluviatilis), 1957–72. J. Fish Res. Board Can. 34:1734–1747.

Hrbáček, J. 1962. Species composition and the amount of zooplankton in relation to the fish stock. Rozpr. Cesk. Akad. Ved Rada Mat. Prir. Ved 72:1–116.

Hurlbert, S. H. and M. S. Mulla. 1981. Impacts of mosquito fish (Gambusia affinis) predation on plankton communities. Hydrobiologia 83:125–151.

Hutchinson, B. P. 1971. The effect of fish predation on the zooplankton of ten Adirondack lakes, with particular reference to the alewife, Alosa pseudoharengus. Trans. Amer. Fish. Soc. 100:325–335.

Hutchinson, G. E. 1957. A treatise on limnology, vol. II. New York: John Wiley and Sons. 1115 pp.

Hutchinson, G. E. 1973. Eutrophication: The scientific background of a contemporary practical problem. Amer. Sci. 61:269–279.

Janus, L. L. and R. A. Vollenweider. 1981. Summary report. The OECD cooperation program on eutrophication. Canadian contribution. Scientific Series No. 131. National Water Research Institute, Inland Waters Directorate, Canada Centre for Inland Waters, Burlington, Ontario.

Jones, J. R. and R. W. Bachmann. 1976. Prediction of phosphorus and chlorophyll levels in lakes. J. Wat. Poll. Control Fed. 48:2176–2184.

Jones, J. R. and M. V. Moyer. 1981. Sportfish harvest predicted as a function of summer standing crop in midwest lakes. Contrib. Missouri Agric. Exper. Station, J. Series 8939.

Kerfoot, W. C. 1987. Cascading effects and indirect pathways, 57–70. In: Predation: Direct and indirect impacts on aquatic communities, Kerfoot, W. C. and A. Sih, eds. Hanover: University Press of New England.

Kitchell, J. A. and J. F. Kitchell. 1980. Size-selective predation, light transmission, and oxygen stratification: Evidence from the recent sediments of manipulated lakes. Limnol. Oceanogr. 25:389–402.

Kitchell, J. F. and S. R. Carpenter. 1987. Piscivores, planktivores, fossils, and phorbins. *In*: Predation: Direct and indirect impacts on aquatic communities, Kerfoot, W. C. and A. Sih, eds. Hanover: University Press of New England.

Liang, Y., J. Melack, and J. Wang. 1981. Primary production and fish yields in Chinese ponds and lakes. Trans. Amer. Fish. Soc. 110:346–350.

Lynch, M. 1979. Predation, competition and zooplankton community structure: an experimental study. Limnol. Oceanogr. 24:253–272.

Lynch, M. and J. Shapiro. 1981. Predation, enrichment and phytoplankton community structure. Limnol. Oceanogr. 26:86–102.

Melack, J. M. 1976. Primary productivity and fish yields in tropical lakes. Trans. Amer. Fish. Soc. 105:575–580.

McCauley, S. and J. Kalff. 1981. Empirical relationships between phytoplankton and zooplankton biomass in lakes. Can. J. Fish. Aquat. Sci. 38:458–463.

McQueen, D. J., and J. R. Post. 1985. Effects of planktivorous fish on zooplankton, phytoplankton, and water chemistry. Lake and Reservoir Management, Proceedings of the Fourth Annual Conference, NALMS, October 1984, McAfee, New Jersey.

McQueen, D. J., J. R. Post, and E. L. Mills. 1986. Trophic relationships in freshwater pelagic ecosystems. Can. J. Fish. Aquat. Sci. 43:1571–1581.

Mills, E. L., J. L. Forney, M. D. Clady, and W. R. Schaffner. 1978. Oneida Lake. *In*: J. A. Bloomfield, ed. Lakes of New York State, vol. II, 367–451. New York: Academic Press.

Mills, E. L. and J. L. Forney. 1983. Impact on *Daphnia pulex* of predation by young perch in Oneida Lake, New York. Trans. Am. Fish. Soc. 112:154–161.

Mills, E. L. and A. Schiavone. 1982. Evaluation of fish communities through assessment of zooplankton populations and measures of lake productivity. N. Am. J. Fish Man. 2:14–27.

Mills, E. L., J. L. Forney, and K. J. Wagner. 1987. Fish predation and its cascading effect on the Oneida Lake food chain. *In*: Predation: direct and indirect impacts on aquatic communities, Kerfoot W. C. and A. Sih, eds., 118–131. Hanover: University Press of New England.

Neilsen, L. A. 1980. Effect of walleye (*Stizostedion vitreum vitreum*) predation on juvenile mortality and recruitment of yellow perch (*Perca flavescens*) in Oneida Lake, New York. Can. J. Fish. Aquat. Sci. 37:11–19.

O'Brien, W. L. and F. DeNoyelles. 1974. Relationship between nutrient concentration, phytoplankton density, and zooplankton density in nutrient enriched experimental ponds. Hydrobiologia 44:105–125.

Oglesby, R. T. 1977. Relationships of fish yield to lake phytoplankton standing crop, production and morphoedaphic factors. J. Fish. Res. Board Canada 34:2271–2279.

Oglesby, R. T. and W. R. Schaffner. 1978. Phosphorus loadings to lakes and some of their responses. Part II. Regression models of summer phytoplankton standing crops, winter total phosphorus and transparency of New York lakes with known phosphorus loadings. Limnol. Oceanogr. 23:135–145.

Pace, M. L. 1984. Zooplankton community structure, but not biomass, influences the phosphorus–chlorophyll relationship. Can. J. Fish. Aquat. Sci. 41:1089–1096.

Paine, R. T. 1980. Food webs, linkage interaction strength, and community infrastructure. J. Anim. Ecol. 49:667–685.

Porter, K. G. 1977. The plant–animal interface in freshwater ecosystems. Am. Sci. 65:159–170.

Sakamoto, M. 1966. Primary production by phytoplankton community in some Japanese lakes and its dependence on lake depth. Arch. Hydrobiol. 62:1–28.

Scavia, D., G. L. Fahnenstiel, M. S. Evans, D. Jude, and J. T. Lehman. 1986. Influence of salmonid predation and weather on long-term water quality trends in Lake Michigan. Can. J. Fish. Aquat. Sci. 43:435–443.

Scott, W. B. and W. J. Christie. 1963. The invasion of the lower Great Lakes by the white perch, *Roccus americanus* (Gamelin). J. Fish. Res. Board Can. 20:1189–1195.

Shapiro, J., B. Forsberg, V. Lamarra, G. Lindmark, M. Lynch, B. Smeltzer, and G. Zoto. 1982. Experiments and experiences in biomanipulation studies of biological ways to reduce algal abundance and eliminate blue-greens. EPA-600/3-82-096. Corvallis Environmental Research Laboratory, U.S. Environmental Protection Agency, Corvallis, Oregon.

Shapiro, J. and D. I. Wright. 1984. Lake restoration by manipulation. Round Lake, Minnesota—the first two years. Freshwater Biol. 14:371–383.

Stenson, J. A. E. 1972. Fish predation effects on species composition of the zooplankton community of eight small forest lakes. Rep. Inst. Freshwater Res. Drottningholm 52:132–148.

Stenson, J. A. E. 1976. Significance of predator influence on composition of *Bosmina* sp. populations. Limnol. Oceanogr. 21:814–822.

Vijverberg, J. and W. L. T. van Densen. 1984. The role of the fish in the food web of Tjeukemeer, The Netherlands. Verh. Int. Ver. Limnol. 22:891–896.

Wagner, K. J. 1983. The impact of natural phytoplankton assemblages on *Daphnia pulex* reproduction. M.S. thesis. Ithaca: Cornell University.

Werner, E. E. and D. J. Hall. 1974. Optimal foraging and the size selection of prey by the bluegill sunfish (*Lepomis macrochirus*). Ecology 55:1042–1052.

Werner, E. E., G. G. Mittelbach, D. J. Hall and J. F. Gilliam. 1983. Experimental tests of optimal habitat use in fish: the role of relative habitat profitability. Ecology 64:1525–1539.

Wetzel, R. G. 1983. Limnology. Saunders, Philadelphia. 743 pp.

CHAPTER 3

Complex Interactions in Oligotrophic Lake Food Webs: Responses to Nutrient Enrichment

William E. Neill

Complex or indirect interactions can be defined as sequences of biotic interactions that functionally link components (species, size classes, functional groups, trophic levels, etc.) in a community. Most aquatic community ecologists think of trophic web interactions among species or size groups in the context of complex interactions, but other indirect linkages also occur through habitats, phenologies, and life cycles. Complex interactions are usually recognized out of the myriad of potential interactions when a perturbation applied to one component is transmitted via the linkages to other components.

From his experimental studies in rocky intertidal systems, Paine (1980) has argued that a relatively few, strong direct and indirect interactions regulate most species' densities and produce community structure. According to this view, even relatively small variations in these regulatory processes would be expected to induce substantial (linear or nonlinear) changes in linked components. Most aquatic ecologists probably subscribe to similar views for many lake communities (Zaret 1980). In contrast, Murdoch and Bence (1987) suggest that most populations exist in regions of community space where strong interactions are minimal and populations are rather loosely coupled. In this latter view, strong direct and indirect interactions are destabilizing and only occur as limiting or boundary conditions at extremes of densities. They are nearly always nonlinear and are detected only with large perturbations. Small perturbations to components fail to be conducted to other components.

Sorting between these two views in any particular community is technically difficult, but can be approximated by experimental perturbations applied in graded steps ranging from near mean to extreme. The shape of the response curve is indicative, with little or no response occurring until extreme densities in Murdoch and Bence's (1987) conceptualization. Unfortunately, much of natural and experimental aquatic community ecology has been heavy handed and qualitative (e.g., fish vs no fish) rather than graded and quantitative. Consequently, it is not often possible to evaluate the relative importance of tightly regulated vs decoupled, density vague processes (Strong 1984).

The following example is typical of our current qualitative understanding of complex interactions in lakes. Formerly fishless lakes inevitably show drastic changes in size and taxonomic composition of zooplankton communities after planktivorous fish introductions. The availability of small bodied prey is in turn widely believed to favor secondary, often invertebrate, predators that are less successful as predators on the larger bodied forms. Further indirect effects may occur through altered grazer–phytoplankton relations (Mills et al. this volume). This familiar sequence of interactions of course is Dodson's (1970) complementary niche concept for the maintenance of invertebrate predators. It is a qualitative model and has been used as such by Zaret (1980) to establish alternative community states. No one, to my knowledge, has examined intermediate conditions between states to assess their validity or describe the nature of transitions. Strongly density dependent processes are usually assumed, but decoupled, density vague relationships are possible as well.

Nonlinear changes in rates of direct and indirect interactions, whether of density dependent, density vague, or abiotic origin, have the potential for generating considerable topography in the landscape of community space. Regions of community space can occur in which different sets of direct and complex linkages predominate. Boundaries, perhaps repelling ones, exist between these multiple domains (Holling 1973), seemingly forming ridges in the landscape between regions of different structure and function. Oksanen et al. (1981) have proposed one such landscape of relatively stable community configurations along a gradient of primary productivity in which communities are separated by ridges of sharp transition associated with establishment of higher trophic levels (Persson et al. this volume). My research has focused on the locations of any such boundary conditions, the identity and strength of (direct and indirect) stabilizing forces within or near them, and the processes generating the breakdown of existing feedback control at boundary conditions. I wish to be able to make predictions of community responses to change of various magnitudes. The theoretical and management implications of such behaviors are obvious.

Small Oligotrophic Lakes

Small lacustrine ecosystems, like rocky intertidal systems, may be good places to study complex interactions. Certain direct interactions such as size selective fish predation upon zooplankton have long histories of study in small lakes, and the evidence for large direct impacts, although not without critics, is widely accepted (Hall et al. 1976). Detecting complex interactions requires tracking a signal generated by a change in some component of the community as it spreads to other parts of the community.

It may be much easier to track such signals in small lakes than in large lakes where signal attenuation through spatial heterogeneity, conflicting disturbances, and distant (if not diffuse) physical boundaries increase the noise component. Limnetic communities in small lakes may also be more strongly influenced by recognizable interactions with littoral and benthic habitats than large lakes (Wetzel 1983), thereby enlarging the scope for finding decipherable trophic, spatial, phenological, and life cycle interactions.

Oligotrophy itself may promote strong direct and indirect interactions, at least at lower trophic levels (potential nutrient competition, limited scope

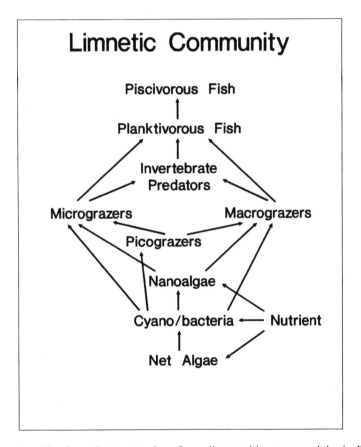

FIGURE 3.1. The limnetic community of an oligotrophic montane lake in British Columbia. Names represent aggregative descriptors of the components of the community and may contain many taxa. The following definitions apply: macrograzers—crustaceans > 0.8 mm; micrograzers—crustaceans, rotifers, and protozoans 5 μm–0.8 mm; picograzers—protozoans < 4 μm; net algae—phytoplankton > 80 μm; nannoalgae—largely phytoflagellates 3–8 μm; cyanobacteria—myxotrophic blue-greens and heterotrophic bacteria < 3 μm.

for adaptive responses to grazing or predation, etc.). In addition, small oligotrophic lakes are highly responsive to rate manipulations, particularly via nutrient enhancement, so linked interactions can frequently be identified and measured within the time and space scales imposed by limited funding. Enclosure based studies are particularly well suited to oligotrophic conditions, as edge effects due to periphyton growth seem less troublesome. In short, small oligotrophic lakes allow efficient experimental manipulations of species' linkages, rate structures, and habitat contexts.

Figure 3.1 shows the general structure of a planktonic food web in small oligotrophic lakes in montane coastal British Columbia. In any particular situation, some of these linkages are very much more significant than others and the resulting community may differ substantially from one in a different context. Fish, for example, are absent in many lakes and ponds because of geographic barriers to dispersal. I am interested in: (1) defining the processes, especially including complex interactions, producing various community structures among these montane lakes and (2) defining the limits of control by those processes, i.e., uncovering any boundary conditions between various community configurations. Physical, chemical, and biological processes undoubtedly influence (probably nonlinearly) the balance of biotic rate processes in producing one structure or another. I have employed mesocosm and whole lake manipulations of some of these processes to begin to identify and evaluate their relative impacts. I will relate three sets of results from these studies that illustrate methodological and organizational insights particularly well. Two of the studies have been published, while the third is still underway and details are less certain. As this paper is intended to discuss some of the broader perspectives, I will minimize data in most of the following discussion.

Primary Productivity and the Dominance of Herbivory vs Invertebrate Planktivory

In naturally fishless Gwendoline Lake, zooplankton community structure and dynamics are largely controlled by exploitative interactions within and among macrograzers and micrograzers (Fig. 3.1). Evidence of strong food limitation exists from all summers except one during the past 14 yr. The top predator, discounting sparse salamander larvae, is the large larval phantom midge *Chaoborus trivittatus*. Its high first instar mortality, however, prevents it from attaining sufficient density to suppress many crustacean species (Neill 1981). Field surveys in the geographic region indicate the existence of very few oligotrophic lakes and ponds in which *Chaoborus* plays an important organizing role (Neill unpublished data). However, both the literature on *Chaoborus* in more productive environments and my own qualitative experiments using high nutrient loadings indicated that zooplankton community structure can be strongly regulated by this large opportunistic predatory insect under eutrophic conditions (Neill and Peacock 1980). By using a logarithmically graded series of nutrient additions at natural N/P ratios (> 35:1 by atoms) in large enclosures, I hoped to locate the transition boundary between herbivore vs planktivore domination of the community, identify the factor(s) responsible for *Chaoborus*' recruitment failure and success on opposite sides of the boundary, and evaluate whether tight direct or indirect linkages between components occurred only at the boundary (Murdoch and Bence 1987) or at intermediate densities as well (Paine 1980).

In Gwendoline Lake, first instar *Chaoborus* consume micrograzers, namely calanoid copepod nauplii and solitary rotifers, both of which are scarce when first instar larvae appear. Starvation or starvation mediated cannibalism were hypothesized as the agents of mortality, and the boundary between poor and good recruitment was hypothesized to depend on rotifer or naupliar density. In a series of logarithmically graded nutrient treatments, no increases in the densities of rotifers or *Chaoborus* larvae were detected under any nutrient loadings within the oligotrophic range given by Wetzel (1983), despite increased productivity of the existing chrysophyte–cryptophyte algal community. Only at nutrient loadings > 5× natural inputs to Gwendoline Lake (i.e., at the oligotrophic–mesotrophic transition) were sufficient rotifers produced to enhance *Chaoborus* recruitment. Rapidly increasing impacts of *Chaoborus* predation on the crustacean community were detected under mesotrophic conditions (Neill 1984, 1985) until devastating effects were seen under eutrophy (Neill and Peacock 1980). Thus, a boundary between minimal *Chaoborus* impact and rapidly increasing impacts on the prey community occurs near the oligotrophy–mesotrophy transition and is caused by relief of the bottleneck in juvenile *Chaoborus* recruitment. The approach to the boundary is flat (i.e., without detectable rotifer response) and the boundary is at some considerable distance from natural, oligotrophic algal concentrations. Consequently, it is unlikely to be frequently crossed by normally observed variations in phytoplankton production in these oligotrophic lakes.

An Alternative Pathway to
Dominance by *Chaoborus*

The near absence of rotifers from Gwendoline Lake over most of the oligotrophic nutrient loading range was unexpected, because rotifers are widely reported from other oligotrophic systems. Dramatic *Daphnia rosea* density responses to artificially enhanced primary productivity, however, provided the clue that competition by the largest cladocerans in these systems (2 mm maximum length) might suppress the micrograzers. In fact, when *Daphnia* were experimentally excluded from two fertilized enclosures, very minute quantities of added nutrients sufficed to produce dramatic rotifer blooms (Neill 1984).

Under natural oligotrophic conditions, both complete removal and logarithmically graded removal of *Daphnia rosea* revealed this large species' preeminence in excluding rotifers during the summer. Complete removal of *Daphnia* reduced daily grazing rates by 70%, and enhanced algal biomass more than two times and rotifer abundance 150-fold (Neill 1984, 1985). Both exploitative and interference competition (Gilbert and Stemberger 1984) are probably involved in rotifer suppression. In any case, *Chaoborus* survivorship increased more than fivefold after rotifers bloomed in *Daphnia*'s absence. Subsequent predatory impacts by third and fourth instar *Chaoborus* upon the remaining crustaceans were very large (Neill 1985), with small forms like *Bosmina*, *Ceriodaphnia*, *Diaphanosoma*, and copepod nauplii being reduced up to 85% of control densities. *Diaptomus* spp. were also suppressed as *Chaoborus* became a strong regulator of most crustacean species. *Chaoborus*' control of the community under oligotrophic conditions could thus be achieved merely by elimination of the dominant herbivore *Daphnia*.

Daphnia competition suppression of rotifers (and so, *Chaoborus*) is the density dependent agent that subdivides the low primary productivity region into alternative community configurations. However, from the above total removal experiment, it is still unclear how strong *Daphnia*'s control may be. A pulsed, graded removal of *Daphnia* (simulating spring recruitment failures) was used to measure *Daphnia*'s ability to regain its control. Graded removal of *Daphnia* revealed that *Daphnia* could still exclude rotifers despite springtime recruitment failures of 60%. At least 90% recruitment failure of *Daphnia* was necessary to allow abundant rotifers in this zooplankton community (Neill 1985). Thus, *Daphnia*'s indirect control of rotifers through food competition indirectly controlled *Chaoborus* over a very broad range of *Daphnia* densities. Very substantial recruitment failures or large increases in algal productivity are necessary to overwhelm *Daphnia*'s regulatory role in the community. Such strong regulation at normal densities is much more characteristic of Paine's (1980) view of population and community organization than Murdoch and Bence's (1987).

Whether large perturbations to primary productivity or *Daphnia* recruitment sufficient to allow rotifer escape occur with any detectable frequency in nature is still unknown. Whether enhanced densities of *Chaoborus trivittatus* could be maintained indefinitely in this new community configuration by preying heavily upon all crustaceans (including *Daphnia*) is also uncertain. A small proportion of oligotrophic montane ponds and lakes that I have sampled in the Pacific Northwest are seemingly dominated by *Chaoborus* and may represent an alternative stable state (*sensu* Connell and Sousa 1983). *Chaoborus* are several times more dense in these systems, species abundances and mean body sizes are substantially different, and cyclomorphic changes in morphology of some taxa are evident, suggesting enhanced *Chaoborus* predation. Rotifers are also considerably more dense than in *Daphnia* dominated lakes, potentially allowing large numbers of *Chaoborus* to recruit successfully. An experimental test of the stability of this rare configuration involving a whole lake enhancement of *Chaoborus trivittatus* is in the offing.

Daphnia and Fish Planktivores

The above studies indicate that invertebrate predation by *Chaoborus* can substantially affect the zooplankton community only in certain regions of community space: that is, above oligotrophic levels of nanoflagellate productivity or at densities of *Daphnia* well below resource determined maxima. Intense fish predation, of course, has a suppressing effect on *Daphnia* abundance, at least in mesotrophic and eutrophic lakes (Hall et al. 1976). In unproductive systems, however, it is unclear just how much of an impact sparse resident fish populations at carrying capacity may have on the biomass of such crustaceans. Certainly the literature is very mixed on the subject. Further, after an experimental introduction of trout and char to oligotrophic fishless lakes within a few km of Gwendoline Lake, we were surprised to find no detectable effects on the biomasses of dominant cladocerans or copepods, despite some reduction in mean body sizes (Northcote et al. 1978; Walters et al. 1987). Only minor increases in fish tolerant invertebrate predators such as cyclopoid copepods occurred. Food limitation of these predatory invertebrate taxa again seems to be the principal constraint (Peacock 1982).

In nearby oligotrophic salmon producing lakes, on the other hand, anadromous fish can have much larger impacts than resident salmonids, especially during quadrennial peak years. Adults that have grown to maturity at sea not only import large numbers of young to these lakes, but also heavy nutrient subsidies (from parental carcasses) that are crucial in supporting the high densities of fry during peak years (Hyatt and Stockner 1985). Thus, huge pulses of planktivorous fry and nutrients every 4 yr make these oligotrophic lakes a special case of exceptional planktivory. That *Daphnia* are absent in these systems and rotifers are abundant suggests that fish predation under oligotrophy (if sufficiently intense) may function similarly to recruitment failure (of *Daphnia*) in fishless lakes. On the other hand, Stockner et al. (this volume) suggest that inappropriate particulate food for these macrograzers, not intense fish predation, is the principal proximate factor in *Daphnia*'s absence.

Algal Composition and
Zooplankton Community Structure

From the above experiments in fishless lakes, it is apparent that strong, direct effects of *Daphnia* grazing define the nannophytoplankton densities under which invertebrate predation by *Chaoborus* can structure the com-

munity. Breakdown of *Daphnia*'s control of algal abundance indirectly
permits *Chaoborus* to become dominant. What would happen, if, instead
of cryptophytes and chrysophytes, the phytoplankton was altered to other
forms? Would *Daphnia* still predominate and continue to almost exclude
very small grazers and indirectly exclude invertebrate predators? Evidence
from marine ecosystems suggests that major changes in grazer and predator
composition may stem directly from bottom–up changes in phytoplankton,
viz, flagellates vs diatoms (Greve and Parsons 1977; Parsons 1979). To
effect a change in phytoplankton composition, fertilizer at a low *N:P* ratio
(ca 12:1) was directly added to Gwendoline Lake to determine if altered
algal quality would drive the plankton community across other unknown
organizational boundaries. Could this community be perturbed into a new
configuration by redirecting energy and nutrients from the bottom of the
food web through alternative pathways to favor different predatory forms?

Gwendoline Lake was modestly fertilized every week during the ice
free period in 1979 with a low N:P fertilizer mixture. A heavy phytoplankton bloom of desmids and thick-walled or gelatinous green algae appeared within 3 wk, but was apparently consumed by the dominant grazer
Daphnia rosea, which rapidly increased to about $10\times$ its historical maximum abundance. By the following spring, however, slightly more abundant calanoid copepod nauplii and *Daphnia* neonates were the only evidence of the previous year's fertilization. Algal composition had returned
to nannoplanktonic phytoflagellates and all grazer densities were within
historical ranges by midsummer. Invertebrate predators realized no extra
benefit from the nutrient supplementation, as few grazers besides *Daphnia*
were capable of using the coarse algal taxa that had been enhanced. Thus,
Daphnia maintained its dominance with this alteration of the phytoplankton. This result further illustrates *Daphnia*'s strong regulatory role in the
plankton community in fishless Gwendoline Lake.

About 1 km downstream from the outlet of Gwendoline Lake is a similar
oligotrophic lake (Eunice Lake) which contained an identical zooplankton
assemblage, save one diaptomid copepod and the absence of *Chaoborus*
predators. Introduced *Salmo clarki* had eliminated the *Chaoborus* some
years earlier, and their small stable population relied heavily upon terrestrial insects as their chief food. Their effects on plankton biomass have
been undetectably small, suggesting that they are not heavy planktivores.
None of the resident invertebrate predators had filled in for *Chaoborus*,
a result of food limitation of their juvenile stages (Peacock 1982). Eunice
Lake received less than 5% of the low N/P nutrients added to Gwendoline
(Marmorek 1982). Nevertheless, in marked contrast to Gwendoline Lake,
relatively minor fertilization of Eunice Lake produced large blooms of
picoplankton, mainly < 2 μm cyanobacteria, probably *Synechococcus*,
and large trichomes (> 150 μm) of *Anabaena* and *Oscillatoria*. Resident
chrysophytes and cryptophytes initially became very scarce. The phytoplankton thus changed from mainly 8–20 μm particles to very small and

very large particles. *Daphnia rosea* showed very low fertility despite an abundance of tiny cells, and its population declined to less than 10% of historical densities. *Daphnia rosea*'s use of picoplankton is inefficient (DeMott and Kerfoot 1982; Neill unpublished data) and may even be deleterious (Arnold 1971). Perhaps interference with feeding occurred in the presence of blue-green trichomes (Webster and Peters 1978). Within a month of the rise in picoplankton, very small phyto- and zooflagellates 2–5 μm in length appeared in numbers up to 400× historical values. Rotifers proliferated and nauplii of calanoid and previously rare cyclopoid copepods survived extremely well, probably consuming flagellates (Peacock 1982), although they may also have used picoplankton directly. Other small crustacean grazers also increased 25–400%. Fertility of all these taxa was several times higher than historical values, presumably in response to increased food supplies. By late summer, total zooplankton biomass was elevated by about 50% over the historical mean, but composition was very much more diverse, especially among protozoans, rotifers, small cladocerans, and predatory cyclopoid copepods. Although very few new species were seen in the lake and no species were lost, the entire character of the community had changed. Relative abundances of species had changed dramatically and the size spectrum was shifted to decidedly smaller organisms.

By the next spring, long after nutrient inputs had returned to their normal range, adult and copepodite stages of the predatory copepod *Diacyclops thomasi* were more abundant than any other crustacean taxon (14% of the total biomass vs < 0.1% in previous years). Over the next 12 months, these vigorous predators sequentially eliminated nearly every small crustacean and rotifer species, as well as the young of all large taxa. By late June, 2 yr after fertilization, *Diacyclops* adults, copepodites, and nauplii comprised 99% of the biomass, the remainder being made up of *Bosmina*, *Chydorus*, and a few miscellaneous taxa. Cyanobacteria no longer dominated the phytoplankton, as more typical cryptophytes and chrysophytes had reestablished. Nevertheless, no *Daphnia*, *Diaptomus*, *Diaphanosoma*, or *Ceriodaphnia* were recorded in any sample for over 18 months, while *Diacyclops* persisted in very high numbers by cannibalizing its own herbivorous nauplii and copepodites. This bizarre community eventually collapsed, probably under the weight of its own economics. Recruitment from resting eggs or upstream Gwendoline Lake gradually restored the community, with *Diacyclops* stabilizing at about 3% of the biomass 4 yr after fertilization.

Clearly, a very major boundary had been crossed as a result of a seemingly inconsequential level of enrichment at an atypical nutrient ratio. Apparently, traditional energy flow pathways from intermediate-sized algae to *Daphnia* were constricted by the change and previously competitively inferior pathways involving the "microbial loop" (Azam et al. 1983) were opened. *Daphnia* was unable to respond to these changes and better adapted small grazers exploited the abundant new resources. Food limited cyclopoid predators capitalized on the abundance of small grazers and quickly dominated.

Summary

The aforementioned manipulative probes have identified three regions of community space in which different sets of processes dominate in these small oligotrophic lakes. Relatively sharp boundaries separate these regions, in each of which is a structurally rather different configuration of taxa that has its own set of complex interactions and some degree of robustness. In all cases, one or two direct interactions and a few indirect interactions could be identified inside each region. As judged from graded perturbations of the communities inside these regions, these interactions appear to limit abundances of most species over a considerable range of variation in conditions. When perturbations were strong, boundaries were encountered in which the dominant strong interactions relatively rapidly failed to control the effects of alternative strong interactions, and the dominant processes in community structure changed. Figure 3.2 summarizes the regions identified to date.

The predominant configuration is the historical community in a region of low nutrient loadings and high N/P ratios. Highly grazable chrysophytes

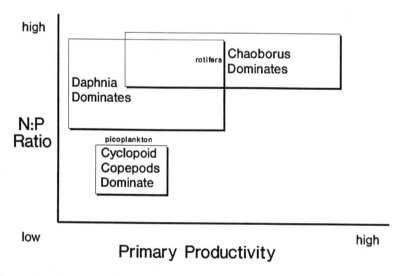

FIGURE 3.2. Summary of regions of Gwendoline Lake community space dominated by different ecological processes. Conditions involved in the boundaries between regions are indicated, viz, the abundances of solitary rotifers and the abundance of picoplankton relative to nanoplankton.

and cryptophytes are exploited by a grazer community strongly dominated by *Daphnia*. This region is quite robust to natural and experimental variation in primary production and phytoplankton composition, mainly because of *Daphnia*'s broad diet and strong, nonlinear functional responses to available food. The limits of *Daphnia*'s domination are reached when its total predatory response to grazable algal biomass approaches saturation or when nannoplankton are replaced by picoplankton and/or net plankton.

The second major configuration exists in the same nutrient ratio–productivity region as the previous community, but is controlled by the invertebrate predator *Chaoborus trivittatus*. Very large perturbations to *Daphnia*'s control (large nutrient additions or large recruitment failures) are required to get into this domain from the first configuration via a complex, indirect route. It is still unclear whether *Chaoborus* can sustain this new configuration indefinitely through its heavy predation. The small number of oligotrophic lakes that I have found in this configuration suggest either that it may be unstable and transient, or that it is merely difficult to escape *Daphnia*'s control.

The third region identified to date is one that also occurs under relatively low nutrient conditions, but only when input N/P ratios are low. Absence of *Chaoborus* may be a further requirement. Under low N/P stimulation, a community of mainly picoplanktonic cyanobacteria, hetero/myxotrophic flagellates, rotifers, and small crustaceans develops. Picoplanktonic cyanobacteria seem to develop under other conditions as well, and low N/P ratios may not be a requirement in these lakes. In any case, the community configuration of life-forms may be unstable because it is highly invasible by predatory cyclopoid copepods. Heavy planktivory by fish (Porter et al. this volume) or *Chaoborus* may be able to control the cyclopoid predation and prevent self-destruction, but this layer of complexity is as yet experimentally untested. Interestingly, the boundary between this community configuration and the historical community seems quite easy to cross, given the proper nutrient competition among primary producers. Very small nutrient additions at low N/P ratios sufficed to swing the balance. Twenty times heavier fertilizations with this low N/P nutrient mix favored coarse algal forms that were consumed by the dominant herbivore *Daphnia*, and no reordering of the plankton community occurred. The nutrient ratio, loading rate, and pulse frequency may be involved in configuring the algal species assemblage.

In the oligotrophic lakes that I have been studying, complex interactions of a variety of sorts occur within recognizable community configurations. Reminiscent of rocky intertidal systems, a few strong interactions occur in each configuration of organisms and their effects spread into the web via indirect linkages. Components of the web are then functionally connected in a particular manner. Boundaries between these configurations seem to occur as a consequence of loss of control by the dominant direct interactions, the impacts of which change over a small range of conditions.

Strong interactions occur at both high (planktivore) and low (nannoplankton and picoplankton) trophic levels, with their effects being transmitted through indirect linkages up and down the food web. It seems that future study designs must consider potentially important direct effects occurring at any level of the community and possibly spreading in any direction. By using graded manipulations of various magnitudes, it should be possible to define the ranges over which particular strong interactions can dominate and to locate the limits at which their control begins to be superseded.

Concluding Remarks

Ideas in aquatic ecology have a tendency to get lost in technical issues of measurement or in low-level processes having only vague potential relation to populations or communities. To be certain, there are good reasons that our science has a technical flavor. Nevertheless, aquatic ecology desperately needs methods, viewpoints, and theories that will expand horizons beyond the details of "my lake" or "my organism(s)." Strategies such as graded perturbations are needed to expand our measurements beyond the *status quo*, especially when grant duration is so short. Strategies of how to compare the organization of aquatic communities or where to fit them in relation to each other are needed to make comprehensible the blizzard of detail. Alternative ways of viewing the components of ecological systems may be especially useful, particularly as size and taxonomy can give only partial perspectives. Finally, a willingness to tolerate qualitative perspectives on the gross scale as well as quantitative relationships on the fine scale may prove extraordinarily helpful in developing conceptual frameworks. The study of complex interactions, while providing an opportunity for another layer of reductionistic extremism in explaining "my lake," also provides considerable opportunity for synthesis, generality, and integration. In the course of our mad-cap races for Least Publishable Units, I hope that the synthetic importance of complex interactions is not drowned in a sea of special cases.

Acknowledgements. Funding for the research discussed in this paper was provided by the Natural Sciences and Engineering Research Council of Canada and by The University of British Columbia. Helpful comments on the manuscript by S. R. Carpenter and J. Kitchell are much appreciated.

References

Arnold, D. E. 1971. Ingestion, survival and reproduction by *Daphnia pulex* fed seven species of blue-green algae. Limnol. Oceanogr. 16:906–920.
Azam, F., T. Fenchel, J. E. Field, J. S. Gray, L. A. Meyer-Reil, and F. Thingstad. 1983. The ecological role of water column microbes in the sea. Mar. Ecol. Prog. Ser. 10:257–263.

Connell, J. H. and W. P. Sousa. 1983. On the evidence needed to judge ecological stability or persistence. Am. Nat. 121:789–824.

DeMott, W. R. and W. C. Kerfoot. 1982. Competition among cladocerans: the interaction between *Bosmina* and *Daphnia*. Ecology 63:1949–1966.

Dodson, S. I. 1970. Complementary feeding niches sustained by size-selective predation. Limnol. Oceanogr. 15:131–137.

Dodson, S. I. 1974. Adaptive changes in plankton morphology in response to size-selective predation: a new hypothesis. Limnol. Oceanogr. 19:721–729.

Gilbert, J. and R. Stemberger. 1984. Control of *Keratella* populations by interference competition by *Daphnia*. Limnol. Oceanogr. 30:180–188.

Greve, W. and T. R. Parsons. 1977. Photosynthesis and fish predation: hypothetical effects of climate change and pollution. Helgolander wiss. Meersunters 30:666–672.

Hall, D. J., S. T. Threlkeld, C. W. Burns, and P. H. Crowley. 1976. The size–efficiency hypothesis and the size structure of zooplankton communities. Ann. Rev. Ecol. Syst. 7:177–208.

Holling, C. S. 1973. Resilience and stability of ecological systems. Ann. Rev. Ecol. Syst. 4:1–24.

Hyatt, K. D. and J. G. Stockner. 1985. Responses of sockeye salmon (*Onchorynchus nerka*) to fertilization of British Columbia coastal lakes. Can. J. Fish. Aq. Sci. 42:320–331.

Marmorek, D. 1983. The effect of lake acidification on zooplankton community structure and phytoplankton–zooplankton interactions: an experimental approach. M.Sc. Thesis. The University of British Columbia.

Murdoch, W. W. and J. Bence. 1987. General predators and unstable prey populations. *In*: Predation: direct and indirect impacts on aquatic communities, ed. W. C. Kerfoot and A. Sih, 17–30. Hanover: University Press of New England.

Neill, W. E. 1981. Impact of *Chaoborus* predation on the structure and dynamics of a crustacean zooplankton community. Oecologia, Berlin. 48:164–177.

Neill, W. E. 1984. Regulation of rotifer densities by crustacean zooplankton in an oligotrophic lake in British Columbia. Oecologia, Berlin. 61:175–181.

Neill, W. E. 1985. The effects of herbivore competition upon the dynamics of *Chaoborus* predation. Arch. Hydrobiol. Beih. 21:483–491.

Neill, W. E. and A. Peacock. 1980. Breaking the bottleneck: Interactions of nutrients and invertebrate predators in oligotrophic lakes. *In*: Ecology and evolution of zooplankton communities, ed. W. C. Kerfoot. Hanover: University Press of New England.

Oksanen, L., S. D. Fretwell, J. Arruda, and P. Niemela. 1981. Exploitation ecosystems in gradients of primary productivity. Am. Nat. 118:240–261.

Paine, R. T. 1980. Food webs: linkage, interaction strength and community infrastructure. J. Anim. Ecol. 49:667–686.

Parsons, T. R. 1979. Some ecological, experimental and evolutionary aspects of the upwelling ecosystem. S. Afr. J. Sci. 75:536–540.

Peacock, A. 1982. Responses of *Cyclops bicuspidatus thomasi* to alterations in food and predators. Can. J. Zool. 60:1446–1462.

Strong, D. R. 1986. Density vagueness: Adding the variance in the demography of real populations. *In*: Community Ecology, ed. J. Diamond and T. J. Case, New York: Harper and Row.

Walters, C. J., E. Krause, W. E. Neill, and T. G. Northcote. 1987. Equilibrium models for seasonal dynamics of plankton biomass in four oligotrophic lakes. Can. J. Fish. Aq. Sci. In press.

Webster, K. E. and R. H. Peters. 1978. Some size-dependent inhibitions of larger
    cladoceran filters in filamentous suspensions. Limnol. Oceanogr. 23:1238–1244.
Wetzel, R. 1983. Limnology. 2d ed. Philadelphia: W. B. Saunders.
Zaret, T. M. 1980. Predation and freshwater communities. New Haven: Yale Uni-
    versity Press.

CHAPTER 4

Predator Regulation and Primary Production Along the Productivity Gradient of Temperate Lake Ecosystems

Lennart Persson, Gunnar Andersson,
Stellan F. Hamrin, and Lars Johansson

Introduction

More than any other ecological discipline, limnology has claimed to be an ecosystem oriented and holistic science (Rodhe 1979; Wetzel 1983). In other ecological disciplines, an intense debate has taken place recently concerning whether or not the ecosystem is a meaningful ecological unit (e.g., Simberloff 1980 vs Levins and Lewontin 1980), but this has not been the case in limnology. We will argue that, although the ecosystem orientation in limnology is not the same as that represented by systems ecology today (Patten and Odum 1981; Odum and Biever 1984), both have a common historical root. Consideration of this historical circumstance is essential for understanding why the importance of complex intertrophic interactions has largely been neglected in limnology. The development of limnology has been treated by Elster (1974) and Rigler (1975), among others. We take a somewhat different approach by focusing mainly on what has constrained the development of limnology. In doing so we consider two factors which have had major impacts on the discipline: (1) lake typology, and (2) the multidisciplinary nature of limnology. A third factor, the integration between theory and application, has strongly influenced the development of the discipline but will not be treated in this context.

This chapter presents an alternative approach to the study of lake ecosystems. Our conceptual framework is based on the notion that the dynamics of lake ecosystems can be understood only through an integrated view of the ecosystem including regulatory mechanisms at different trophic levels. Other researchers have started to do this (Benndorf et al. 1984; Carpenter et al. 1985; Kerfoot 1987; Kitchell and Carpenter 1987; Mills et al. 1987; Threlkeld 1987). Our contribution is to treat ecosystem regulation in relation to a productivity gradient (see also Riemann et al. 1986). We also suggest that the discussion of whether trophic level biomasses are controlled from below or from above is superficial to some extent, and can be dissolved by viewing bottom–up and top–down factors as parts of an integrated process. The approach is based on theoretical developments in general ecology (Oksanen et al. 1981). Empirical data are presented to support the theory, although a broader data set is needed before a closer evaluation of the theory is possible.

Holistic Claims in Limnology

The holistic approach in limnology can be traced back to Forbes' famous paper, "The lake as a microcosm" (1887). Holism here is linked with the concept of a balance of nature which was also a part of the antique holism.

. . . life does not perish in the lake, nor even oscillate to any considerable degree, but on the contrary the little community secluded here is as prosperous as if its state were one of profound and perpetual peace. (Forbes 1887, p. 549)

In Forbes' version the older idea of God as a supervisor of the balance has been replaced by Nature as an almost divine abstraction (". . . Nature has turned upon these favored children. . . .") (Forbes 1887, p. 545). This use of the term Nature can be related to the problems Forbes and his contemporaries had in explaining the relative order in nature despite the fact that individuals, which were part of this nature, were engaged in a hard struggle against each other for survival. In a later paper, Forbes' concept of balance is underlined by his view of the natural harmony between predator and prey and the deleterious effects of rapid population variations (Forbes 1903). Forbes' major contribution to the development of ecology is as a predecessor of the superorganismic school (Clements 1916). This view of the ecosystem later influenced the development of systems ecology, where cybernetics have been introduced to explain the stability of ecosystems (Odum 1969; Patten and Odum 1981; Odum and Biever 1984).

Forbes' superorganismic view influenced both Naumann and Thienemann. This is particularly evident in Thienemann's works e.g.,

Every life community forms together with the environment which it fills, a unity and often a community so closed in itself that it must be called an organism of the highest order (vide Bodenheimer 1958).

An explicit superorganismic view can be found even in quite recent papers (Rodhe 1979; Forsberg 1982). The main impact on limnology of the superorganismic view has been to form the basis for the development of lake typology, because one consequence of the superorganismic view is that lakes are viewed as "individuals." Lake typology has played a prominent role in limnology (Naumann 1919, 1932; Thienemann 1920, 1921; Elster 1958; Rodhe 1969, 1975). Lake typologists viewed lakes as static systems (Rigler 1975), and consequently emphasized differences rather than similarities between lakes and, more important, were reluctant to accept the transformation of one system into another. Lake typology has been criticized (Rodhe 1975), but its continued influence is illustrated by the high frequency of case studies in limnology and the viewpoint that each lake is unique. Moreover, this criticism has not considered that the classification of lakes was based on an Aristotelian view of nature as composed of ideal types (Simberloff 1980), a view challenged by the Darwinian revolution (Lewontin 1974).

Another basis for limnology's claims to holism is its multidisciplinary nature, including subdisciplines such as chemistry, hydrology, ecology, taxonomy, and geology (Rodhe 1979; Wetzel 1983). However, the different subdisciplines have not traditionally had the same priority, and there has been a strong bias towards studies of physical and chemical conditions of lakes. When biotic interactions are considered, they tend to be concentrated at lower trophic levels. Even though this bias has been questioned for more than 2 decades, there is still a strong emphasis on studies of physical and chemical conditions. The bias is well illustrated by the scanty treatment of fish in limnological textbooks (Table 4.1), which is most pronounced in earlier textbooks (Welch 1952; Ruttner 1963; Hutchinson 1957–1975). Recent limnology texts (Goldman and Horne 1983; Wetzel 1983) admit that fish may have significant regulatory effects on other trophic levels: "Their impact on the operation of the system in terms of carbon flux and nutrient regeneration at times can be quite significant. . . ." (Wetzel 1983, p. 658). However, the indirect effects of fish predation on primary producers and water chemistry are essentially ignored. An exception is Moss (1980), who devotes 18% of his book to fish and, more important, also discusses the indirect effects of fish predation.

One explanation for the omission of higher trophic levels from limnology is that experimental nutrient additions generally cause an increase in phytobiomass (Schindler 1974; Gerhart and Likens 1975; O'Brien and De-

TABLE 4.1. Numbers and proportion of pages dealing with fish in relation to the total number of pages in limnological textbooks

	Numbers		Proportion
	Total	Fish	%
Ruttner (1963) Fundamentals of Limnology	279	26 words	– –
Welch (1952) Limnology	442	~ 2	0.5
Hutchinson (1957–1975) Treatise on Limnology	2467	~ 5	0.2
Wetzel (1975) Limnology	660	~ 11	1.7
Wetzel (1983) Limnology	752	~ 16	2.1
Moss (1980) Ecology of Fresh Waters	293	~ 52	18
Goldman & Horne (1983) Limnology	433	~ 23	5.3
Likens (ed.) (1985) An Ecosystem Approach to Aquatic Ecology; Mirror Lake and its Environment	444	~ 16	3.6

noyelles 1976; but see Riemann and Søndergaard 1986). This might also explain why the concept of limiting factors has played a more prominent role in limnology than in other ecological disciplines (Rigler 1975). Another reason for neglecting the role of higher trophic levels is the significant relationships between phosphorus loading levels and phytobiomass (Dillon and Rigler 1974; Nichols and Dillon 1978; Schindler 1978; Watson and Kalff 1981) and between phytoplankton and zooplankton biomass (McCauley and Kalff 1981; Mills and Schiavone 1982; Canfield and Watkins 1984; Hanson and Peters 1984) that have been found for lakes of different trophic conditions. These studies show that substantial variation in biomass of lower trophic levels is statistically explainable by nutrient supply (Fig. 4.1). The significant regression between phosphorous loading and phytobiomass has in turn formed the basis for models predicting the highest allowable nutrient loading of lakes (Vollenweider 1968). Regression analyses have also considered higher trophic levels from a bottom–up perspective (Oglesby 1977; Hanson and Leggett 1982; Mills and Schiavone 1982).

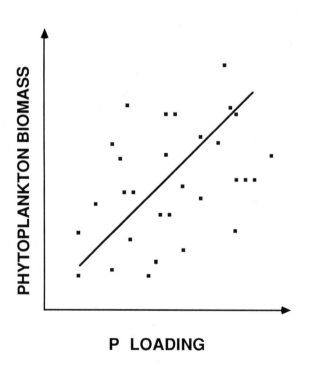

P LOADING

FIGURE 4.1. A schematic picture of the phosphorus–phytoplankton relationship as viewed by many limnologists (Dillon and Rigler 1974; Nichols and Dillon 1978; Schindler 1978; Watson and Kalff 1981).

Experimental Challenges to the Chemical Bias in Limnology

Although a significant regression between P and phytoplankton has been found (Fig. 4.1), substantial variance in phytoplankton biomass remains unexplained (Shapiro 1978; Carpenter and Kitchell 1987). We argue that, while nutrient loading is clearly important in explaining variation in phytoplankton biomass, the dynamics of aquatic ecosystems cannot be understood by a simple regression approach. Rather, the trophic structure of ecosystems is another variable that significantly affects ecosystem dynamics. By considering this variable, the phosphorous–phytoplankton biomass relationship turns out to be more complex than what is suggested from Figure 4.1.

Hrbáček et al. (1961) provided an early demonstration that phytoplankton biomass is not a simple reflection of nutrient loading. They showed that the presence of fish markedly decreased zooplankton biomass and

TABLE 4.2. Experiments testing the effect of fish on lower trophic levels performed in lakes (1), reservoirs (2), ponds (3), enclosures (4), or aquaria (5). The results are given as being positive ($+$), indifferent (0), or negative ($-$) to the hypothesis that zooplanktivorous fish reduce the biomass of herbivorous zooplankton (Z), decrease water transparency and/or increase the biomass of phytoplankton (Ph/T), and increase the nutrient content of the water (N). Lack of sign indicates absence of information

	Fish	Z	Ph/T	N	
1	Cyprinidae	+	+		Hrbáček et al. 1961
1	*Alosa pseudoharengus*	+	+		Brooks and Dodson 1965
1	Cyprinidae	+	+	+	Stenson et al. 1978
1	Cyprinidae	+	+	+	Henrikson et al. 1980
1	Coregonidae, *Perca fluviatilis*	+	+		Reinertsen and Olsen 1984
2	Cyprinidae	+	+		Leah et al. 1980
2	*Gasterosteus* sp.	+	+		Olrik et al. 1984
3	Cyprinidae	+	0		Grygierek et al. 1966
3	*Lepomis macrochirus*	+	0		Hall et al. 1970
3	*Gambusia gambusia*	+	+		Hurlbert et al. 1972
3	Cyprinidae	+	+		Losos and Hetesa 1973
3	Cyprinidae	+	+		Fott et al. 1980
3	*Gambusia gambusia*	+	+		Hurlbert and Mulla 1981
3	*Pimephales* sp.	+	+		Spencer and King 1984
3	*Lepomis macrochirus*	+	+	+	Hambright et al. 1986
4	Cyprinidae			+	Lamarra 1975
4	Cyprinidae	+	+	+	Andersson et al. 1978
4	*Lepomis macrochirus* *Pimephales* sp.	+			Lynch 1979
4	*Lepomis macrochirus*	+	+		Lynch and Shapiro 1981
5	*Pimephales* sp.	+	+		Elliott et al. 1983

increased phytoplankton biomass. Subsequently, numerous experiments have shown that introductions or removals of primary carnivores (i.e., planktivorous fish) have marked effects on lower trophic levels (Table 4.2). Generally, these experiments have yielded the same qualitative results irrespective of experimental system (aquarium, pond, lake enclosure, or whole-lake experiments) or species of fish (Table 4.2). Differences have been found. For example, cyprinid fishes have stronger effects than percid fishes (Andersson 1984; Persson 1987). Apart from the effects on plankton, nutrient conditions have also been affected. Moreover, it has also been shown in some cases that the presence of planktivorous fish may be a prerequisite for an increase in phytoplankton biomass after nutrient addition (Riemann and Søndergaard 1986).

While experimental studies have consistently demonstrated strong effects of fish on lower trophic levels (Table 4.2), the implications of these results are still open to debate among limnologists. The experimentally derived results have been predicted by theory (Rosenzweig 1971, 1973; Wollkind 1976; Rosenzweig and Schaffer 1978). Furthermore, the theoretical models predict a wide range of possible outcomes with respect to stability properties of the systems, predictions that remain to be tested.

An Alternative Holistic View

The multidisciplinary nature of limnology has resulted in a bias towards studies of chemistry and lower trophic levels. This bias in turn fails to account for important interactions in lake ecosystems, as the experimental studies discussed above have shown. The other holistic claim in limnology, the superorganismic view, has been heavily criticized from different directions (Gleason 1926; Simberloff 1980; Levins and Lewontin 1980). Levins and Lewontin have characterized the superorganismic view of ecosystems as being a holistic idealism, where the different populations are defined by the community properties; hence there is a one to one relation between the community and the populations. In contrast to this, Levins and Lewontin propose that the properties of the community and its populations are linked by what they call many to one and one to many relations; i.e., the whole and the part do not completely determine each other. The first notion implies that there are many possible configurations of populations which preserve the same qualitative properties at the community level. One implication is that the different populations cannot be defined simply by knowledge of the community properties. The second notion implies that not all properties of the parts (populations) are specified by rules at the level of parts. A given part of the community could consist of different populations solving a certain problem in different ways. Consequently, a change in some parameter (abiotic or biotic) within the community affects the populations differently depending upon both the population properties and the structure of the community. One such important

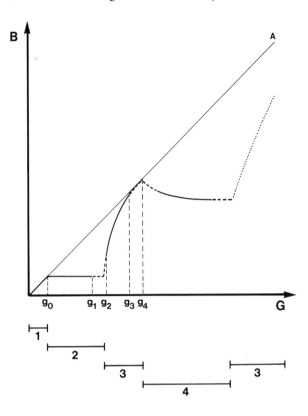

NUMBER OF TROPHIC LEVELS

FIGURE 4.2. The relationship between potential primary productivity (G) and phytobiomass (B) for a one to four trophic-level system based on Oksanen et al. (1981). Line A shows the relationship between G and B in the absence of other trophic levels. For very low G values phytobiomass will increase linearly with G up to a level where phytobiomass is high enough to maintain grazers in the system ($G < g_0$). A further increase in G will give no response in phytobiomass but only an increase in grazer biomass ($g_0 - g_1$). If G continues to increase, in systems with a high ecological efficiency, the system may become destabilized so that sustained oscillations are to be expected ($g_1 - g_2$). A further increase in G will lead to the primary carnivore isocline being passed and an unstable plant–grazer–primary producer equilibrium point is established. With a further increase in G this equilibrium point will be stabilized ($g_2 - g_3$). In aquatic systems with a high ecological efficiency a fourth trophic level—secondary carnivores—might be able to invade the system ($G > g_4$). The effects of a secondary carnivore is to regulate primary carnivores, and an increase in G in these systems, therefore, should result only in an increase in biomass of secondary carnivores and herbivores and not in primary carnivores and phytobiomass. In fact, phytobiomass is predicted to decline somewhat, especially at the transition from three- to four-link dynamical structure. ------ denotes our suggested relation between G and B in high-productivity lakes (not formally analyzed).

structure within a community is the number of trophic levels, which according to Levins and Lewontin determines whether a certain level is controlled by predation or competition. As a result, a nutrient input will affect either the biomass or the turnover rate and age distribution.

An argument similar to that of Levins and Lewontin (1980) has recently been put forth by Oksanen (1988). In contrast to Odum (Odum 1969; Patten and Odum 1981; Odum and Biever 1984) he argues that a purely Darwinian theory of ecosystem structure can explain ecosystem order (disorder) and stability (instability) and that no cybernetic and mutualistic controls (ultimately superorganismic qualities) are needed to understand ecosystem structure. The work of Oksanen (1988; see also Fretwell 1977; Oksanen 1983; Oksanen et al. 1981) is thus an attempt to develop an evolutionary ecosystem ecology rivaling the previously dominant superorganismic system ecology developed by Odum and coworkers. The graphical model presented by Oksanen et al. (1981) considers trophic-level interactions along a productivity gradient (Fig. 4.2). Based on this model, a general framework to understand ecosystem regulation of temperate lake ecosystems may be developed. A first attempt at this was presented by Riemann et al. (1986), and the remaining part of our paper is a more complete development of this perspective.

A Graphical Model for Trophic-Level Interactions in Relation to Primary Productivity

The experimental studies discussed above considered the effects of planktivorous and/or benthivorous fish on lower trophic levels. During recent years experiments dealing with systems including a fourth trophic level-piscivores-have been performed (Shapiro and Wright 1984; Benndorf et al. 1984; Hambright et al. 1986; Carpenter et al. 1987). Briefly, these experiments show that addition of secondary carnivores to three-level systems may cause a decrease in planktivorous fish and phytoplankton and an increase in zooplankton. These results were predicted by theoretical models for four-level systems developed by Smith (1969), although they were not recognized by his contemporary limnological colleagues, and more completely articulated by Oksanen et al. (1981; see also Fretwell 1977 for a first verbal version).

Oksanen et al. (1981) analyzed how enrichment of the environment affects trophic-level interactions and trophic chain lengths (Fig. 4.2), thereby dissolving the superficial contradiction between whether abiotic or biotic factors determine the properties of ecosystems. Instead the impacts of these two factors are regarded as one integrated process. From this follows a rejection of the view that trophic-level biomasses are controlled either from above or from below (cf. McQueen et al. 1986). The primary production allowed by the system in absence of higher trophic levels, G, is the independent variable (Fig. 4.2) and depends on available resources (nutrients, light availability). The model is based on equilibrium conditions, and we know that environmental variation characterizes most natural systems. However, we view the model as a useful starting point for further elaboration, and recent theoretical developments have introduced the effects of seasonal variation (Oksanen pers. comm.). Hence, although a basic assumption of the model may be violated in many systems, the model should be evaluated according to its contribution in understanding ecosystem structure. We also feel that the integration of bottom-up or top-down factors is a decisive advantage compared to other attempts in this research area.

The model makes the interesting prediction that for two- and four-level systems no positive relation between G and phytobiomass exists, while such a relationship is present in the three-level system (Fig. 4.2). Furthermore, the model predicts that food chain length is dependent on primary productivity. The latter prediction is supported by the observations that in nonhumic Swedish lakes with very low productivity and a high transparency (> 7 m), i.e. with a very low phytobiomass, pelagic piscivores are absent (Fig. 4.3). The data also suggest that piscivores invade the benthic/littoral habitat before the pelagic habitat and that the change in the proportion of total piscivores is largely explained by a change in the proportion of piscivorous perch (*Perca fluviatilis*). According to Oksanen et al. (1981), G and phytoplankton biomass should be uncoupled in four-level systems, i.e. in lakes where piscivores form a substantial biomass. Studies from four low to medium productivity lakes in southern Sweden, each with a high percentage of piscivores, support this prediction, because although primary production differed among lakes by three times or more, the phytoplankton biomass was the same (Table 4.3).

In four-level systems the proportion of piscivorous fish in the total fish biomass is high, and so is the ratio of zooplankton to phytoplankton biomass (Fig. 4.4A). Our data are derived from south Scandinavian lakes, and their applicability to temperate lakes on other continents is not known. Interestingly, however, one of the most investigated lakes in the USA, Mirror Lake, has trophic structure and biomass (including bacteria) similar to those predicted from its primary production using data from European lakes (Fig. 4.4A; Likens 1985, p. 342; Riemann et al. 1986, p. 271). Thus, although the fish communities consist of quite different species, the trophic structure appears to be the same.

So far we have considered systems with a primary production less than 70 mg C m^{-2} yr^{-1}. With a progression to highly productive lakes, our data show a major decrease in the proportion of piscivorous fish to less than 10% of total fish biomass (Fig. 4.3, Fig. 4.4B). This is a result of a drastic increase in the biomass of cyprinids (Persson 1983b, 1983c; Johansson and Persson 1986). That piscivores (primarily piscivorous perch, Fig. 4.3) do not respond to the increased biomass of cyprinids is primarily related

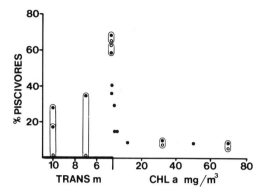

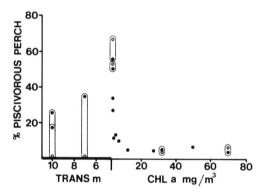

FIGURE 4.3. Percentage of piscivores in total fish biomass in relation to phyto-plankton biomass or lake transparency. Upper panel: total piscivores (including perch, pike *Esox lucius*, burbot *Lota lota*, and sander *Stizostedion lucioperca*). Lower panel: piscivorous perch only. A chlorophyll *a* content of ~ 3 mg m^3 corresponds to a transparency of 4–5 m in these nonhumic lakes; hence, the X-axis is continuous although the scale changes at a transparency of 4.5 m. In lakes where both benthic and pelagic habitats were sampled (O), data for benthic (●) and pelagic (○) habitats are presented separately. Study lakes (*n* = 15) are given in appendix.

to the competitive interaction between perch and cyprinids (primarily roach, *Rutilus rutilus*) (Persson 1983a; Johansson and Persson 1986). In very productive lakes with decreased habitat heterogeneity, perch face increased competition from cyprinids, and both factors negatively affect the recruitment of perch to piscivorous stages (Persson 1983a, 1986). The increase of cyprinids, especially roach, with productivity is related to an increase in the biomass of blue-green algae, exploited only by roach, and to a relative change in the abundance of animal food resources favoring cyprinids in the competition with perch (Persson 1983b, 1983c, 1986, 1987;

TABLE 4.3. Phytoplankton primary production and mean biomass and transparency in four Swedish lakes (see Appendix) with piscivore fish exceeding 35% of the total fish biomass

Lake	Primary production (g C m^{-2} yr^{-1})	Phytoplankton biomass (mg chlorophyll a m^{-3})	Transparency (m)
Kalvsjon	20	2.0	4–5
Fegen	20	2.3	4–5
Bolmen	20	5.2	4–5
Ivosjon	60–75	2.6	4–5

Johansson and Persson 1986). Thus, in highly productive lakes, perch populations are skewed towards smaller size classes due to increased interspecific competition, and the percentage of piscivores in the population is dramatically lowered.

The low proportion of piscivores in highly productive lakes means that the four-level system is functionally reduced to a three-level system in

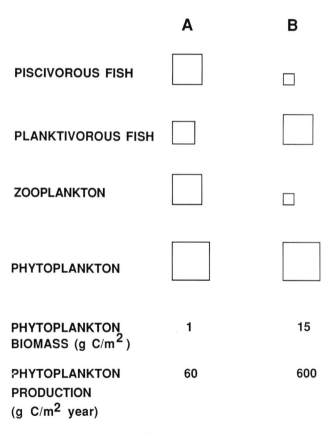

FIGURE 4.4. Relative biomass of different trophic levels (as depicted by different sized squares) in model lakes (see Riemann et al. 1986) with a primary production of 60 g C/m^2 year (*A*) and 600 g C/m^2 year (*B*), respectively.

which primary production and biomass should be positively related. This suggestion is supported by empirical data (Gelin 1975, Gelin and Ripl 1978), but it was not predicted by Oksanen et al. (1981). One reason for this discrepancy between theory and empirical data may be that the assumption of perfectly distinct trophic levels made by the model is violated (i.e. when species undergo ontogenetic niche shifts; Werner and Gilliam 1984). In this case small perch belong to the third trophic level and competition from cyprinids at this stage may limit their recruitment to piscivorous classes (Persson 1983a, 1983c, 1986; Johansson and Persson 1986).

To conclude, the available data on aquatic systems provide tentative support for the predictions of Oksanen et al.'s (1981) model for systems of up to four trophic levels. More studies including lakes from different continents are, however, required to confirm the usefulness and generality of the model. In highly productive systems our data show a return to three functional trophic levels. For this type of system, a theoretical development of the model allowing piscivores to be recruited through the planktivore trophic level is necessary. Evidence from lakes throughout the temperate zone for our suggested relationship between the number of functional trophic levels and productivity comes from the well-known succession pattern of fish communities with productivity (Svärdson 1976; Hartmann and Nümann 1977; Kitchell et al. 1977; Leach et al. 1977; Persson 1983c). This succession pattern from salmonids, over the percids to cyprinids with increasing productivity suggests changes also in number of functional trophic levels.

Both theory and empirical data suggest that along a productivity gradient two zones (two-level and four-level systems) will be found where no positive relationship exists between potential primary production and phytoplankton biomass (Fig. 4.2). If the data from the whole productivity gradient of Figure 4.2 is included, a significant regression of B on G will of course be found in analogy with the significant regression of B on phosphorous loading (Fig. 4.1). However, the underlying trophic interactions are not unraveled by this simple regression. Furthermore, it can be questioned whether interactions between different trophic levels can be analyzed at all with a simple regression approach as used by McQueen et al. (1986). To understand the dynamics of a certain level, both the number of trophic levels in the system and the position of the level in relation to the top one must be considered. The type of response of a lake ecosystem to nutrient supply will therefore also depend on the trophic system present.

Competition and Predation under Different Productivity Conditions

So far we have mainly been concerned with how stability properties and equilibrium densities were affected by the number of trophic levels. Implicit in the model by Oksanen et al. (1981) is the assumption that predator and prey are regulated by different factors. Top carnivores are predicted to be resource-limited, while plants as a trophic level should be controlled by grazers in even-linked systems and controlled by resources in odd-linked systems (Fig. 4.5). Actually the model is an extension of the HSS hypothesis (Hairston, Smith and Slobodkin 1960) to more than three trophic levels. Fretwell (1977) also suggested that an increase in primary production should result in an increase in the food chain length. Our empirical data support this prediction up to a four-level system (Fig. 4.3), but in opposition to the prediction the number of trophic levels will functionally decrease to 3 levels in highly productive systems (Fig. 4.5).

It follows that planktivorous fish are mainly controlled by competition in three-level systems. This inference is supported by data from highly productive lakes where the configuration of the fish community is primarily explained by the competitive superiority of cyprinids (Persson 1983b, 1983c, 1986; Johansson and Persson 1986). In contrast, planktivorous fish in four-level systems are expected to be mainly predator-controlled (Fretwell 1977). The two factors - predation and competition - are expected to have different effects on the size structure of planktivorous fish populations. High competition intensity is expected to result in populations consisting of small individuals due to food shortage (Persson 1983b), whereas high predation pressure on smaller individuals should result in the opposite size distribution. The predictions of predation control in four-level systems

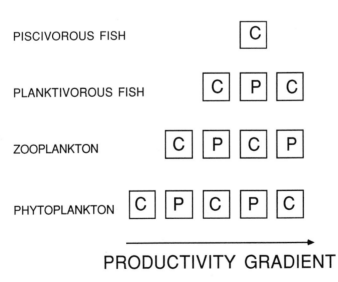

FIGURE 4.5. Number of trophic levels in productivity gradient. In contrast to Fretwell (1977), we suggest that the number of trophic levels is functionally reduced to three in high-productivity systems. The main structuring forces on different trophic levels are also indicated (C = competition, P = predation).

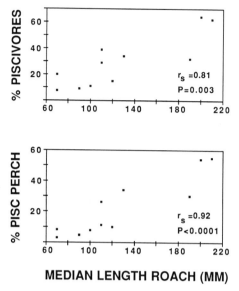

MEDIAN LENGTH ROACH (MM)

FIGURE 4.6. Median size of a planktivorous fish (roach) in relation to the proportion of piscivorous fish in total fish biomass. Upper panel: total piscivores. Lower panel: piscivorous perch. Study lakes ($n = 11$) are given in appendix.

and competition control in three-level systems are corroborated by empirical data, since systems with a high proportion of piscivores (four-level systems) have a higher median size of planktivorous roach than systems with a low proportion of piscivores (three-level systems; Fig. 4.6). Zooplankton and phytoplankton regulation patterns have not been studied within the conceptual framework outlined here. Based on Fig. 4.5, however, phytoplankton should be more resource limited in high than in medium productivity systems (four level systems). The evolutionary consequences of being controlled by different main factors were discussed by Fretwell (1977). For example, grazer- or predator-limited organisms are expected to be short-lived and unmanageable by potential predators, whereas food-limited organisms should be long-lived and may either be poisonous or chemically distasteful to predators.

Concluding Remarks

The conceptual framework outlined above forms a valid base for understanding ecosystem regulation in temperate lakes. It also dissolves the contradiction between abiotic or biotic factors as chief regulators of lake ecosystems, since the impacts of these factors are viewed in an integrated process. The data support the usefulness of this approach, although we are fully aware that more empirical data and experimental tests are necessary, especially from other geographical regions. Although we have not discussed the implications of seasonal variation (Porter 1977), this has, as previously mentioned, been included in recent theoretical developments (Oksanen 1987). To conclude, changes in trophic structure can be advanced as an important complementary explanation for the variation in primary production of lakes not explained by nutrients.

Acknowledgements. We thank Steve Carpenter, Chris Luecke, Gary G. Mittelbach, Lauri Oksanen, Bo Riemann, Staffan Ulfstrand, and an anonymous reviewer for valuable comments on earlier versions of this paper. Support from the National Swedish Environmental Protection Board, the Swedish Natural Science Research Council and the Fisheries Research Board of Sweden is gratefully acknowledged.

Appendix

Lakes

The lakes studied are all situated in Southern Sweden. Detailed data on the lakes are given in Hamrin (1973, 1974, 1984; Lakes Bolmen, Ivösjön and Vombsjön), Collvin and Persson (1975; Lakes Northern and Southern Osbysjön), Lundkvist (1975; Lakes Fegen and Kalven), Hamrin et al. 1977 (Lakes Skärlen and Skärsjön), and Gelin (1979; Lake Öresjö). This information is available on request as mimeographed reports in Swedish from the Department of Ecology, Limnology, Lund, Sweden. Information in international publications is available for Lake Hinnasjön (Andersson et al. 1975) and Lake Sövdeborgssjön (Persson 1983a). Summarized information in international publications concerning most of the lakes are given in Andersson et al. (1975), Gelin (1975), Persson (1983a), and Hamrin (1986a, b).

Fishing Methods

In lakes not stratified during summer, fish composition has been estimated with benthic gill nets (twined nylon; 8 mesh sizes 16-92 mm stretched measure). In stratified lakes, pelagic fishing with other gill nets has also been performed (monofilament nylon; 8 mesh sizes 18-100 mm). Benthic net were placed at different depths to include the total bathymetric range. Pelagic nets were placed in the epi-, meta- and hypolimnion. At least 5 and 6 nets respectively were used during each of 2 or 3 subsequent 24-h periods. The gill nets were set at 9 a.m. and pulled 24 h later. In lakes deeper than 4 m, the catch at each depth has been weighted with respect to its share of the total lake volume. CPUE (catch per net and 24 h) is used as a measure of fish biomass.

References

Andersson, G. 1984. The role of fish in lake ecosystems—and in limnology. *in*: Interaksjoner mellom trofiske nivaer i ferskvann, ed. B. Bosheim and M. Nicholls, 189–197. Nordisk limnolog symposium, 1984, Oslo.

Andersson, G., H. Berggren, and S. F. Hamrin. 1975. Lake Trummen restoration project. III. Zooplankton, macrobenthos and fish. Verh. int. Ver. Limnol. 19: 1097–1106.

Andersson, G., H. Berggren, G. Cronberg, and C. Gelin. 1978. Effects of planktivorous and benthivorous fish on organisms and water chemistry in eutrophic lakes. Hydrobiologia 59: 9–15.

Benndorf, J., H. Kneschke, K. Rossatz, and E. Penz. 1984. Manipulation of the pelagic food web by stocking with predacious fishes. Int. Revue ges. Hydrobiol. 69: 407–428.

Bodenheimer, F. S. 1958. Animal ecology today. Uitgererij, Dr. W. Junk, Den Haag.

Brooks, J. L. and S. I. Dodson. 1965. Predation, body size and composition of plankton. Science 150:28–35.

Canfield, D. E. Jr., and C. E. Watkins. 1984. Relationships between zooplankton abundance and chlorophyll *a* concentration in Florida lakes. J. Freshwat. Ecol. 2: 335–344.

Carpenter, S. R. and J. F. Kitchell. 1987. The temporal scale of variance in limnetic primary production. Am. Nat. 129: 417–433.

Carpenter, S. R., J. F. Kitchell and J. R. Hodgson. 1985. Cascading trophic interactions and lake productivity. BioScience 35:634–639.

Carpenter, S. R., J. F. Kitchell, J. R. Hodgson, P. A. Cochran, J. J. Elser, M. M. Elser, D. M. Lodge, D. Kretchmer, X. He, and C. N. von Ende. 1987. Regulation of lake primary productivity by food-web structure. Ecology 68: 1863–1876.

Clements, F. E. 1916. Plant succession; An analysis of the development of vegetation. Carnegie Institution, Washington.

Collvin, L. and L. Persson. 1975. Lake Osbysjön—A descriptive study with suggestions on restoration measures. Inst. of Limnology, Lund. Mimeographed (in Swedish).

Dillon, P. J. and F. H. Rigler. 1974. The phosphorus-chlorophyll relationship in lakes. Limnol. Oceanogr. 19:767–773.

Elliott, E. T., L. G. Castanares, D. Perlmutter and K. G. Porter. 1983. Trophic-level control of production and nutrient dynamics in an experimental planktonic community. Oikos 41:7–16.

Elster, H.-J. 1958. Das limnologische Seetypensystem, Rückblick und Ausblick. Verh. Internat. Verein. Limnol. 13: 101–120.

Elster, H.-J. 1974. History of Limnology. Mitt. Internat. Verein. Limnol. 20:7–30.

Forbes, S. A. 1887. The lake as a microcosm. Bull. Peoria (Illinois) Sci. Ass. Bull. 77–87; reprinted in Illinois Natur. Hist. Survey Bull., 1925. 15:537-550.

Forbes, S. A. 1903. Studies of the food of birds, insects and fishes. Bull. Ill. State lab. Nat. Hist. 1:3–18.

Forsberg, C. 1982. Limnological research can improve and reduce the cost of monitoring and control water quality. Hydrobiologia 86:143–146.

Fott, J. L., L. Pechar and M. Prazakova. 1980. Fish as a factor controlling water quality in ponds. *in*: Hypertrophic ecosystems. Developments in Hydrobiology 2, ed. L. Mur and J. Barica, 255–261. The Hague: Dr. W. Lunk.

Fretwell, S. D. 1977. The regulation of plant communities by the food chains exploiting them. Perspect. Biol. Med. 20: 169–185.

Gelin, C. 1975. Nutrients, biomass and primary productivity of nannoplankton in eutrophic Lake Vombsjön, Sweden. Oikos 26:121–139.

Gelin, C. 1979. Limnological study of Lake Öresjö. Inst. of Limnology, Lund. Mimeographed (in Swedish).

Gelin, C. and W. Ripl. 1978. Nutrient decrease and response of various phytoplankton size fractions following the restoration of Lake Trummen, Sweden. Arch. Hydrobiol. 81:339–367.

Gerhard, D. Z. and G. E. Likens. 1975. Enrichment experiments for determining nutrient limitation: four methods compared. Limnol. Oceanogr. 20:649–653.

Gleason, H. A. 1926. The individualistic concept of the plant association. Bull. Torrey Bot. Club 56:7–26.

Goldman, C. R. and A. J. Horne. 1983. Limnology. New York: McGraw-Hill.

Grygierek, E., A. Hillbricht-Ilkowska and I. Spodniewska. 1966. The effect of fish on plankton community in ponds. Verh. int. Ver. Limnol. 16:1359–1366.

Hairston, N. G., F. E. Smith and L. B. Slobodkin. 1960. Community structure, population control and competition. Am. Nat. 94:421–425.

Hall, D. J., W. E. Cooper and E. E. Werner. 1970. An experimental approach to the production dynamics and structure of freshwater animal communities. Limnol. Oceanogr. 15:839–928.

Hambright, K. D., R. J. Trebatoski, and R. W. Drenner. 1986. Experimental study of effects of bluegill (*Lepomis macrochirus*) and largemouth bass (*Micropterus salmoides*) on pond community structure. Can. J. Fish. Aquat. Sci. 43: 1171–1176.

Hamrin, S. F. 1973. The bathymetric distribution and relative composition of the fish fauna in the north and south parts of Lake Bolmen 1970–1972. Inst. of Limn., Lund. Mimeographed (in Swedish).

Hamrin, S. F. 1974. Lake Ivösjön—a limnological study 1973. Inst. of Limn., Lund. Mimeographed (in Swedish).

Hamrin, S. F. 1984. The fish community and its food resources in Lake Vombsjön 1983. Inst. of Limn., Lund. Mimeographed (in Swedish).

Hamrin, S. F. 1986a. Ecology of vendace, *Coregonus albula*, with special reference to factors important to the commercial fishery. Arch. Hydrobiol. Beih. 22: 51–72.

Hamrin, S. F. 1986b. Vertical distribution and habitat partitioning between different size classes of vendace, *Coregonus albula*, Can. J. Fish. Aquat. Sci. 43:1617–1625.

Hamrin, S. F., L. Collvin, B. Jonsson, O. Lessmark and L. Persson. 1977. Lake Skärsjön (Huskvarnaån). A limnological study, August 1977. Inst. of Limnology, Lund. Mimeographed (in Swedish).

Hanson, J. M. and W. C. Leggett. 1982. Empirical prediction of fish biomass and yield. Can. J. Fish. Aquat. Sci. 39: 257–263.

Hanson, J. M. and R. H. Peters. 1984. Empirical prediction of crustacean zooplankton biomass and profundal macrobenthos biomass in lakes. Can. J. Fish. Aquat. Sci. 41:439–445.

Hartmann, J. and W. Nümann. 1977. Percids of Lake Constance, a lake undergoing eutrophication. J. Fish. Res. Bd. Can. 34:1670–1677.

Henrikson, L., H. G. Nyman, H. G. Oscarson and J. E. Stenson. 1980. Trophic changes, without changes in the external nutrient loading. Hydrobiologia 68:257–263.

Hrbáček, J., M. Dvorakova, Korinek, V. and L. Prochazkova. 1961. Demonstration of the effect of the fish stock on the species composition and the intensity of the metabolism of the whole plankton association. Verh. Int. Verein. Limnol. 14:192–195.

Hurlbert, S. H. and M. S. Mulla. 1981. Impacts of mosquitofish (*Gambusia affinis*) predation on planktonic communities. Hydrobiologia 83:125–151.

Hurlbert, S. H., J. Zedler and D. Fairbanks. 1972. Ecosystem alteration by mosquitofish (*Gambusia affinis*). Science 175:639–641.

Hutchinson, G. E. 1957. A treatise on limnology. Vol 1. New York: John Wiley and Sons.

Hutchinson, G. E. 1967. A treatise on limnology. Vol. 2. New York: John Wiley and Sons.

Hutchinson, G. E. 1975. A treatise on limnology. Vol. 3. New York: John Wiley and Sons.

Johansson, L. and L. Persson. 1986. The fish community of temperate eutrophic lakes. *in*: Carbon dynamics in eutrophic, temperate lakes, ed. B. Riemann and M. Søndergaard, 237–266. Amsterdam: Elsevier Science Publishers.

Kerfoot, W. C. 1987. Cascading effects and indirect pathways. *in*: Predation: direct and indirect impacts on aquatic communities, ed. W. C. Kerfoot and A. Sih, 57–70. Hanover: University Press of New England.

Kitchell, J. F. and S. R. Carpenter. 1987. Piscivores, planktivores, fossils, and phorbins. *in*: Predation: direct and indirect impacts on aquatic communities, ed. W. C. Kerfoot and A. Sih, 132–146. Hanover: University Press of New England.

Kitchell, J. F., M. G. Johnson, C. K. Minns, K. H. Loftus, L. Greig and C. H. Olver. 1977. Percid habitat: the river analogy. J. Fish Res. Bd. Can. 34:1959–1963.

Lamarra, V. A. Jr. 1975. Digestive activities of carp as a major contributor to the nutrient loading of lakes. Verh. Internat. Verein. Limnol. 19:2461–2468.

Leach, J. H., M. G. Johnson, J. R. M. Kelso, J. Hartmann, W. Nümann and B. Entz. 1977. Responses of percid fishes and their habitats to eutrophication. J. Fish Res. Bd Can. 34:1964–1971.

Leah, R. T., B. Moss and D. E. Forrest. 1980. The role of predation in causing major changes in the limnology of a hyper-eutrophic lake. Int. Revue ges. Hydrobiol. 65: 223–247.

Levins, R. and R. C. Lewontin. 1980. Dialectics and reductionism in ecology. Synthese 43:47–78.

Lewontin, R. C. 1974. The genetic basis of evolutionary change. New York: Columbia University Press.

Likens, G.E. 1985. An ecosystem approach to aquatic ecology, Mirror Lake and its environment. New York: Springer-Verlag.

Losos, B. and J. Hetesa. 1973. The effect of mineral fertilization and of carp fry on the composition and dynamics of plankton. *in*: Hydrobiological Studies, Vol. 3, ed. J. Hrbáček and M. Straskraba, 173–218. Prague: Academia.

Lundquist, I. 1975. Lakes Fegen and Kalvsjön. A limnological study. Inst. of Limn., Lund. Mimeographed (in Swedish).

Lynch, M. 1979. Predation, competition and zooplankton structure: An experimental study. Limnol. Oceanogr. 24:253–272.

Lynch, M. and J. Shapiro. 1981. Predation, enrichment and phytoplankton community structure. Limnol. Oceanogr. 26: 86–102.

McCauley, E. and J. Kalff. 1981. Empirical relationships between phytoplankton and zooplankton biomass in lakes. Can. J. Fish. Aquat. Sci. 38:458–463.

McQueen, D. J., Post, J. R. and E. L. Mills. 1986. Trophic relationships in freshwater pelagic ecosystems. Can. J. Fish. Aquat. Sci. 43:1571–1581.

Mills. E. L. and A. Schiavone Jr. 1982. Evaluation of fish communities through assessment of zooplankton populations and measures of lake productivity. N. Amer. J. Fish. Manage. 2:14–27.

Mills, E. L., J. L. Forney and K. J. Wagner. 1987. Fish predation and its cascading effect on the Oneida lake food chain. *in*: Predation: Direct and indirect impacts on aquatic communities, ed. W. C. Kerfoot and A. Sih, 118–131. Hanover: University Press of New England.

Moss, B. 1980. Ecology of freshwaters. Oxford: Blackwell.

Naumann, E. 1919. Applied limnology. Some theoretical guide lines for an efficient water culture. Kungl. Landtbruks-akad. Handl. och Tidskr. 199–221. (in Swedish).

Naumann, E. 1932. Grundzüge der Regionalen Limnologie. Die Binnengewässer Band XI. 1–176.

Nichols, K. H. and P. J. Dillon. 1978. An evaluation of phosphorous-chlorophyll-phytoplankton relationships for lakes. Int. Revue ges. Hydrobiol. 63:141–154.

O'Brien, W. J. and F. DeNoyelles Jr. 1976. Response of three phytoplankton bioassay techniques in experimental ponds of known limiting nutrient. Hydrobiologia 49:69–76.

Odum, E. P. 1969. The strategy of ecosystem development. Science 164:262–270.

Odum, E. P. and L. J. Biever. 1984. Resource quality, mutualism and energy partitioning in food chains. Am. Nat. 122: 45–52.

Oglesby. R. T. 1977. Relationships of fish yield to lake phytoplankton standing crop, production, and morphoedaphic factors. J. Fish. Res. Board Can. 34:2271–2279.

Oksanen, L. 1983. Trophic exploitation and arctic phytomass patterns. Am. Nat. 122:45–52.

Oksanen, L. 1988. Ecosystem organization: mutualism and cybernetics or plain Darwinian struggle for coexistence? Am. Nat. 131:424–444.

Oksanen, L., S. D. Fretwell, J. Arruda, and P. Niemela. 1981. Exploitation ecosystems in gradients of primary productivity. Am. Nat. 118:240–261.

Olrik, K., S. Lunder and K. Rasmussen. 1984. Interactions between phytoplankton, zooplankton, and fish in the nutrient rich shallow Lake Hjarbaek Fjord, Denmark. Int. Revue ges. Hydrobiol. 69:389–405.

Patten, B. C. and E. P. Odum. 1981. The cybernetic nature of ecosystems. Am. Nat. 118:886–895.

Persson, L. 1983a. Food consumption and competition between age classes in a perch *Perca fluviatilis* population in a shallow eutrophic lake. Oikos 40:197–207.

Persson, L. 1983b. Food consumption and the significance of detritus and algae to intraspecific competition in roach *Rutilus rutilus* in a shallow eutrophic lake. Oikos 41: 118–125.

Persson, L. 1983c. Effects of intra- and interspecific competition on dynamics and size structure of a perch *Perca fluviatilis* and a roach *Rutilus rutilus* population. Oikos 41:126–132.

Persson, L. 1986. Effects of reduced interspecific competition on resource utilization in perch *Perca fluviatilis*. Ecology 67:355–364.

Persson, L. 1987. Effects of habitat and season on competitive interactions between roach (*Rutilus rutilus*) and perch (*Perca fluviatilis*). Oecologia 73:170–177.

Porter, K. G. 1977. The plant-animal interface in freshwater ecosystems. Am. Sci. 65:159–170.

Reinertsen, H. and Y. Olsen. 1984. Effects of fish elimination on the phytoplankton community of a eutrophic lake. Verh. int. Ver. Limnol. 22:649–657.

Riemann, B. and M. Søndergaard. 1986. Regulation of bacterial secondary production in two eutrophic lakes and in experimental enclosures. J. Plankton Res. 8:519–536.

Riemann, B., M. Søndergaard, L. Persson, L. Johansson. 1986. Carbon metabolism and community regulation in eutrophic, temperate lakes. *in*: Carbon dynamics of eutrophic, temperate lakes, ed. B. Riemann and M. Søndergaard, 267–280. Amsterdam: Elsevier Scientific Publishers.

Rigler, F.H. 1975. Nutrient kinetics and the new typology. Verh. int. Ver. Limnol. 19:197–210.

Rodhe, W. 1969. Limnology, social welfare, and Lake Kinneret. Verh. int. Ver. Limnol. 17:40–48.

Rodhe, W. 1975. The SIL founders and our fundament. Verh. int. Ver. Limnol. 19:16–25.

Rodhe, W. 1979. The life of lakes. Arch. Hydrobiol. Beih. 13:5–9.

Rosenzweig, M. I. 1971. Paradox of enrichment: destabilization of exploitation ecosystems in ecological time. Science 171:385–387.

Rosenzweig, M. I. 1973. Exploitation in three trophic levels. Am. Nat. 107:275–294.

Rosenzweig, M. I. and W. M. Schaffer. 1978. Homage to the Red Queen II. Coevolutionary response to enrichment and exploitation ecosystems. Theor. Pop. Biol. 14:158–163.

Ruttner, F. 1963. Fundamentals of Limnology. Univ. of Toronto Press.

Schindler, D.W. 1974. Eutrophication and recovery in experimental lakes: implications for lake management. Science 184:897–898.

Schindler, D.W. 1978. Factors regulating phytoplankton production and standing crop in the world's freshwaters. Limnol. Oceanogr. 23:478–486.

Shapiro, J. 1978. The need for more biology in lake restoration. Contribution no. 183 from the Limnological Research Center, University of Minnesota, Minneapolis, Minnesota.

Shapiro, J. and D. I. Wright. 1984. Lake restoration and biomanipulation: Round Lake, Minnesota, the first two years. Freshwat. Biol. 14:371–383.

Simberloff, D. 1980. A succession of paradigms in ecology: essentialism to materialism and probabilism. Synthese 43: 3–39.

Smith, F.E. 1969. Effects of enrichment in mathematical models. *in*: Eutrophication: causes, consequences, correctives, 631–645. Washington, D.C.: National Academy of Sciences.

Spencer, C. N. and D. L. King. 1984. Role of fish in regulation of plant and animal communities in eutrophic ponds. Can. J. Fish. Aquat. Sci. 41:1851–1855.

Stenson, J. A. E., T. Bohlin, L. Henrikson, B. I. Nilsson, H. G. Nyman, H. G. Oscarson and P. Larsson. 1978. Effects of fish removal from a small lake. Verh. int. Ver. Limnol. 20:794–801.

Svärdson, G. 1976. Interspecific population dominance in fish communities of Scandinavian lakes. Rep. Inst. Freshwat. Res. Drottningholm 56:144–171.

Thienemann, A. 1920. Die Grundlagen der Biocoenotik und Monards faunistische Prinzipien. In Festschr. fur F. Zschokke, Basel, Kober Verlag 4:1–14.

Thienemann, A. 1921. Seetypen. Naturwissenschaften 18: 1–3.

Threlkeld, S. T. 1987. Experimental evaluation of trophic-cascade and nutrient-mediated effects of planktivorous fish on plankton community structure. *in*: Predation: Direct and indirect impacts on aquatic communities, ed. W. C. Kerfoot and A. Sih, 161–173. Hanover: University Press of New England.

Vollenweider, R. A. 1968. The scientific basis of lake and stream pollution with particular references to phosphorous and nitrogen as eutrophication factors. OECD Techn. Rep. DAS/CSI/68.

Watson, S. and J. Kalff. 1981. Relationships between nannoplankton and lake trophic status. Can. J. Fish. Aquat. Sci. 38:960–967.

Welch, P.S. 1952. Limnology. New York: McGraw-Hill.

Werner, E. E. and J. F. Gilliam. 1984. The ontogenetic niche and species interactions in size-structured populations. Ann. Rev. Ecol. Syst. 15:393–425.

Wetzel, R.G. 1975. Limnology. Philadelphia: Saunders.

Wetzel, R.G. 1983. Limnology (2nd edition). Philadelphia: Saunders.

Wollkind, D. J. 1976. Exploitation in three trophic levels: an extension allowing intraspecific carnivore interaction. Am. Nat. 110:431–447.

Part 2
Microbial Links to the Classical Food Web

CHAPTER 5

Microbial Food Webs in Freshwater Planktonic Ecosystems

John G. Stockner and Karen G. Porter

Introduction

The new microbial paradigm that implicates bacteria, small algae, and protozoa in major pathways for carbon flow was first developed for the pelagic food webs of oligotrophic oceans (Pomeroy 1974, 1984), and is now being recognized in lake and river plankton (Porter 1984; Porter et al. 1985; Stockner and Antia 1986; Stockner 1987; Meyer et al. 1985). The microbial component of freshwater food webs consists of picoplankton such as bacteria, coccoid and rod-shaped cyanobacteria and small eukaryotes (Stockner and Antia 1986), heterotrophic microflagellates (Porter et al. 1985; Riemann 1985) and mixotrophic flagellates (Porter et al. 1985; Bird and Kalff 1986), and ciliates (Pace and Orcutt 1981; Porter et al. 1985). Protozoan members of these groups are also active bacterivores in marine and estuarine microbial food webs along with additional groups not found in fresh water.

Indirect links from pico- and protozooplankton to higher trophic levels are provided by microzooplankton such as rotifers and ciliates (Pace and Orcutt 1981; Gates 1984; Gilbert and Bogdan 1984), and macrozooplankton such as cladocerans and copepods (Pace and Orcutt 1981; Porter et al. 1983). These metazoans can then serve as prey for larger vertebrates and invertebrates (Porter et al. 1979, 1985; Stockner and Shortreed 1988). Studies of marine microbial systems provide evidence that the bacterial, algal, and protozoan components of this microbial food web are metabolically active and have the potential for extremely rapid growth and high rates of feeding and excretion (Johannes 1965; Fenchel 1982a, b, 1986; Sherr and Sherr 1983).

It is likely that the key components of this microbial food web were well represented in the primordial plankton, since small prokaryotic and eukaryotic producers and metazoan consumers have probably been in existence for over 2 billion yr (Fig. 5.1). Fecal pellet remains found in deposits as old as the Gunflint Chert (1.9 billion years) (Robbins et al. 1985) suggest that an extensive planktonic food web existed in the world's oceans at

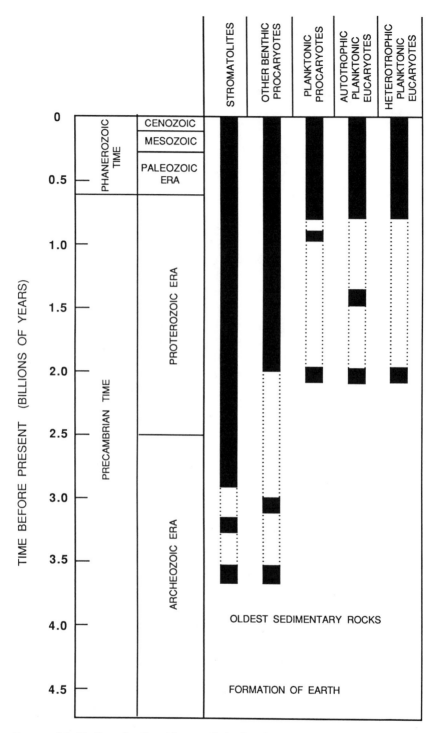

FIGURE 5.1. Earliest fossil evidence of planktonic prokaryotes and eukaryotes. Dashed line denotes incomplete record. (See Robbins et al. 1985).

that time. In fact, the rise of consumers may have led to the rapid diversification of major phytoplankton groups that occurred in the Proterozoic Era.

In this review we summarize the major discoveries that have increased our awareness of microbial processes in freshwater ecosystems, and consider how the pathways of material and energy flow through the microbial community affect the production of carbon and the remineralization of nutrients in the pelagic zone. Though this review will rely largely on examples drawn from studies of plankton communities, we also recognize the importance of similar processes, albeit with a different cast of characters, in the littoral and benthic communities of both lentic and lotic freshwater habitats.

Picoplankton

Populations of heterotrophic picoplankton (bacteria), important in biogeochemical cycles in lakes, were initially considered to be almost exclusively particle associated and were to be found principally in habitats where organic degradation is active, e.g., littoral and benthic sediments (Wood 1965). With the discovery of new staining techniques and epifluorescence microscopy, free-living bacteria were found to be a ubiquitous and abundant component of plankton communities. Abundances range from 10^5 to 10^7 cells/mL with maximum abundance in the epilimnion of temperate lakes during the summer. Recently investigators have shown that bacterial numbers and activity are highly correlated with phytoplankton biomass and production (Porter and Feig 1980; Bird and Kalff 1986; Scavia et al. 1986; Shortreed and Stockner 1986). Measures of bacterial metabolic activity, inferred from tracer studies, were generally quite high, suggestive of populations with the potential for rapid growth and high production (Hobbie et al. 1972; Daley 1979; Riemann and Sondergaard 1986; McDonough et al. 1986). Bacteria can constitute greater than 50% of the combined pico- and nanoplankton biomass in lakes (Caron et al. 1985).

Freshwater autotrophic picoplankton were first reported by Rodhe (1955) from a small lake in northern Sweden, but a decade would pass before Bailey-Watts et al. (1968) would again draw the attention of limnologists to the importance of bacteria sized cyanophytes in Loch Leven, Scotland. By the 1970s and 80s their occurrence had been documented in several European (Chang 1980; Cronberg and Weibull 1981; Craig 1985), African (Melack et al. 1982), North American (Costella et al. 1979; Stockner 1981; Munawar and Fahnenstiel 1982; Craig 1984; Caron et al. 1985; Pick and Caron 1987), and New Zealand lakes (Paerl 1977).

The most ubiquitous algal picoplankters, that are likely a common component of the microbial biomass of nearly all lakes, are the prokaryotic

chroococcoid cyanobacteria *Synechococcus* spp. Small, single- celled eukaryotic picoalgae have also been reported, in one case as being extremely
abundant in acidic, humic-stained (dystrophic) lakes (Stockner 1987). Aspects of the physiology, biochemistry, taxonomy, and ecology of algal
picoplankton have only recently received extensive reviews from both
marine and freshwater systems, and will not be presented in any detail
here (Fogg 1986; Stockner and Antia 1986; Platt and Li 1987).

Algal picoplankton can account for a large fraction of the annual carbon
production, especially in oligotrophic lakes, where 50-70% of the annual
carbon fixation is attributed to organisms passing a 1-2 um filter (Munawar
and Fahnenstiel 1982; Stockner 1988; Stockner and Shortreed 1988). Algal
picoplankton can also contribute substantially to total phytoplankton biomass (chlorophyll), with reported values ranging from 10-45% of the total
chlorophyll (Stockner and Antia 1986). Fahnenstiel et al. (1986) reported
a specific growth rate of 1.5 d^{-1} for algal picoplankton from Lake Superior,
and though the majority of growth measurements have come from marine
cyanobacteria (Stockner and Antia 1986; Platt and Li 1987), the picture
emerging is one of rapid growth and short turnover times. These results
are to be expected because of their minute size and attendant large surface/
volume ratios.

Both algal picoplankton and bacteria are now recognized as key components of plankton communities, especially in oligotrophic lakes, where
they have been shown to be important in material and energy transport,
both seasonally and annually (Stockner 1987; Pick and Caron 1987). Since
the sinking rate of picoplankton is negligible (Beinfang and Takahashi
1983), this source of carbon is not lost to the sediments, but remains in
the epilimnion and is available for consumers. An important question becomes whether or not this source of energy is lost or dissipated by respiration (sink), or is consumed by predators (link) and passed to higher
trophic levels. Despite the known potential for rapid growth of both heterotrophic and autotrophic picoplankton populations, current studies of
their seasonal distribution in the pelagia of lakes does not indicate large
fluctuations in abundance (Porter and Feig 1980; Gude et al. 1985; Pick
and Caron 1987; Scavia et al. 1986; Scavia and Laird 1987; McDonough
et al. 1986; Stockner 1987), suggesting regulation of their numbers by biotic
factors such as grazing. The recent literature provides strong support for
control by predation in both marine and freshwater systems (Azam et al.
1983; Riemann 1985; Stockner and Antia 1986).

Predation on Picoplankton

Small flagellates are presently considered to be the principal grazers of
both bacteria and algal picoplankton (Fenchel 1986; Goldman and Caron
1985; Bird and Kalff 1986). Chroococcoid cyanobacteria have been ob-

served in the food vacuoles of flagellates from the Great Lakes (Caron et al. 1985; Boraas et al. 1985; Fahnenstiel et al. 1985), in coastal British Columbia lakes (Stockner 1987; Stockner and MacIsaac, unpublished data), and in Lake Constance (Gude et al. 1985). Some flagellates appear to be obligate heterotrophs while others tend to be mixotrophic, supplementing their autotrophic nutrition with phagotrophy (Bird and Kalff 1986; Porter 1984; Sanders et al. 1985). The direct utilization of picoplankton by both heterotrophic and autotrophic flagellates provides a tightly coupled link for the transfer of carbon from pico- to nanoplankton sized particles (Sherr et al. 1986; Stockner and Antia 1986). Nanoplankton seem to be more readily grazed by macrozooplankton than are picoplankton (Gliwicz 1980; Porter et al. 1985), and it would appear that predation by flagellates may be an extremely important first step in the transfer of picoplankton carbon to larger metazoans.

Ciliates can be effective bacterivores in aquatic environments, especially in eutrophic lakes where bacterial densities are sufficient to support significant ciliate populations (Fenchel 1984; Porter et al. 1979; Pace and Orcutt 1981; Pace 1982; Gates 1984; Sanders et al. 1985). With their rapid growth potential, ciliates may also play an important role as grazers of small heterotrophic flagellates and pico- and nanophytoplankton, providing yet another link to higher trophic levels. In meso- and eutrophic lakes where bacterial densities are very high, ciliates have been shown to play a key role in energy transfer to higher levels (Porter et al. 1979; Sorokin and Paveljeva 1978).

An annual survey of bacterivory in eutrophic, monomictic Lake Oglethorpe, Georgia, showed that microflagellates dominated grazing at all times of the year (Fig. 5.2). Flagellate grazing impact was 55-99% of the total bacterivory, with mixotrophic flagellates constituting as much as 80% of the total grazing impact during periods of peak bacterial biomass. In late spring to early summer, rotifers and ciliates were responsible for as much as 25 and 30% of the bacterivory, respectively. Grazing impact by crustaceans was generally <1%, with a maximum of 15%. Here, *Ceriodaphnia* and *Daphnia* were the major cladoceran bacterivores (Porter et al. 1983; Pace et al. 1983). Bacterivory by calanoid and cyclopoid copepods was not detected in this study. Total grazing mortality ranged from 11-159% of bacterial production, and averaged 79% throughout the year (Table 5.1). Bacterioplankton in this lake is therefore potentially an important energy and carbon source for higher trophic levels, but primarily via protistan links in the microbial food web.

In oligo- and mesotrophic lakes, rotifers may be the most important link in moving pico- and nanoplankton carbon to macrozooplankton (Stockner and Shortreed 1987; Stockner and MacIsaac, unpublished data). Chroococcoid cyanobacteria have been observed in the guts of rotifers in Lake Ontario (Caron et al. 1985) and in several British Columbia lakes (Stockner 1987). In laboratory studies, rotifers have been shown to graze heavily and grow rapidly on the cyanobacterium *Synechococcus* (Stark-

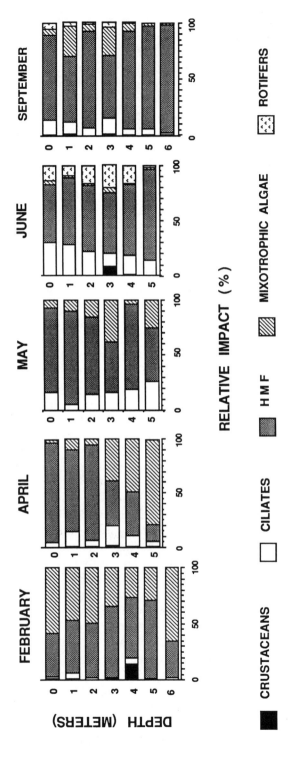

FIGURE 5.2. Relative grazing impact of planktonic bacterivores in Lake Oglethorpe, Georgia. Grazing by cladocerans (solid), ciliates (open), heterotrophic microflagellates (stippled), phagotrophic phytoflagellates (hatched), and rotifers (chevrons) was determined *in situ* using 0.5 μm microspheres as tracers (R. Sanders, K. G. Porter, S. Bennett and A. DeBiase, unpublished data).

TABLE 5.1. Bacterial production and mortality due to grazing in Lake Oglethorpe, Georgia. Numbers in parentheses are standard errors. (From unpublished data of R. Sanders, K. G. Porter, S. Bennet, and A. DeBiase.)

Depth	Standing stock (cells $\times$ 10⁶/ml)	Production	Grazing	% Standing stock removed	Grazing production (%)
		(cells $\times$ 10⁶/ml/day)			
February					
0 m	2.37 (.14)	– –	0.82	35	– –
1 m	2.32 (.67)	– –	0.72	31	– –
2 m	2.28 (.16)	– –	0.74	33	– –
3 m	2.67 (.24)	– –	0.42	16	– –
4 m	3.50 (.22)	– –	0.38	11	– –
5 m	3.20 (.20)	– –	0.23	7	– –
6 m	4.07 (.26)	– –	0.09	2	
April					
0 m	3.90 (.15)	– –	0.45	11	– –[a]
1 m	5.03 (.27)	– –	0.20	4	– –
2 m	6.14 (.29)	– –	0.44	7	– –
3 m	8.70 (.39)	– –	0.22	2	– –
4 m	6.10 (.30)	– –	0.28	4	– –
5 m	7.22 (.29)	– –	0.24	3	– –
May					
0 m	4.30 (.23)	0.16	0.18	4	113
1 m	5.26 (.32)	0.39	0.63	12	162
2 m	4.44 (.26)	0.34	0.35	8	104
3 m	7.23 (.40)	0.36	0.39	5	108
4 m	7.01 (.30)	0.35	0.52	7	148
5 m[b]	5.89 (.37)	0.28	0.18	3	64
June					
0 m	5.95 (.28)	0.20	0.19	3	94
1 m	3.92 (.19)	1.65	0.18	4	11
2 m	5.24 (.24)	2.06	0.46	9	22
3 m	6.65 (.35)	1.52	0.37	6	24
4 m	5.68 (.31)	0.61	0.38	7	63
5 m[b]	6.21 (.27)	0.48	0.76	12	159
September					
0 m	4.40 (.24)	0.31	0.27	4	86
1 m	5.28 (.26)	1.35	0.47	5	35
2 m	5.64 (.24)	1.50	1.33	7	88
3 m	6.26 (.35)	1.59	0.51	6	32
4 m[b]	6.13 (.22)	1.75	0.85	6	49
5 m[b]	7.54 (.33)	1.26	0.73	8	58
6 m[b]	7.13 (.40)	0.44	0.43	7	98

[a]Grazing estimates for April ranged from 55–122% of production measured in April 1985.
[b]Anoxic.

weather, personal communication). Rotifers can sometimes be the most ubiquitous and abundant component of the microzooplankton community, where they characteristically exhibit very high metabolic rates and can often be more abundant than crustaceans (Pace and Orcutt 1981; Gilbert and Bogdan 1984). They appear to be effective grazers of particles in the pico- and small nanoplankton size range, and can exist at threshold food densities commonly found in oligotrophic lakes (Gilbert and Bogdan 1984; Stemberger and Gilbert 1985a, b).

Recently, investigators have shown that the freshwater calanoid copepod *Diaptomus pallidus* can graze effectively on rotifers and shows markedly increased survival and reproduction when rotifers are added to a threshold algal diet (Williamson and Butler 1986). Rotifers appear to graze ciliates effectively in lakes as well (Stenson 1985). This makes them potentially important links between picoplankton and protozooplankton and higher trophic levels.

Freshwater cladocerans, such as *Ceriodaphnia* and *Daphnia*, appear to be able to utilize particles in the pico- and small nanoplankton sizes, albeit not as efficiently as larger nano- and small microplankton sizes (Peterson et al. 1978; Pace et al. 1983; Porter et al. 1983; Porter 1984). *Synechococcus* has been seen in the gut of *Eubosmina* in British Columbia lakes (Stockner 1987), but did not appear digested, an observation similar to that seen for *Synechococcus* in the guts of copepods in the marine environment (Johnson and Sieburth 1982). Perhaps prokaryotic picoplankton cannot be assimilated by larger macrozooplankton, and in passage through the gut they may acquire nutrients for later use (Porter 1973). As in the eutrophic lake system reported above, it seems that most calanoid copepods in oligotrophic lakes cannot effectively graze on picoplankton or very small nanoplankton particles (Scavia and Laird 1987). Therefore, it may well be that cladocerans such as *Daphnia* are the only large macrozooplankton that can effectively utilize all components of the microbial community, ranging in size from 1-50 um.

The presence (or absence) of cladocerans in lakes of varying trophy may markedly affect the degree of activity of microbial metabolism because of top down control (Scavia and Fahnenstiel this volume; Crowder et al., this volume). This is apparent from studies of British Columbian coastal and interior lakes. The presence of *Daphnia* as a keystone species in the food webs of interior lakes results in high efficiency of transfer of pico- and nanophytoplankton production to higher trophic levels, which include juvenile sockeye salmon (Stockner and Shortreed 1988) (Fig. 5.3). In coastal ultra-oligotrophic lakes without *Daphnia*, the flow of carbon is from picoplankton to microflagellates, ciliates, rotifers and copepods. The additional length of these extended food webs provides less efficient carbon flow, resulting in small-bodied, low-biomass populations of juvenile sockeye salmon in these lakes (Porter et al. this volume).

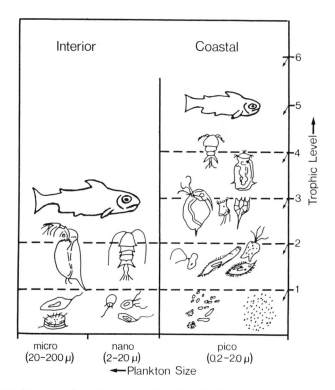

FIGURE 5.3. Nanoplankton-based food chains of oligotrophic interior British Columbia lakes are shorter and more efficient because of the presence of *Daphnia*, producing larger individuals and supporting higher standing stocks of planktivorous juvenile sockeye salmon. The food chains of ultra-oligotrophic coastal lakes are picoplankton based with rotifers, copepods, and *Eubosmina* as predominant zooplankters. The food webs tend to be more complex, longer, and less efficient, supporting lower standing stocks of juvenile sockeye of a smaller average size (Stockner and Shortreed 1988).

The Microbial Loop

The microbial processes and pathways for energy flow involved in the return of dissolved organic matter (DOM) from phytoplankton exudation or leakage to the conventional or classic (phytoplankton-zooplankton-fish) food chain were termed the microbial loop (Azam et al. 1983). The key components of the loop include heterotrophic and autotrophic pico- and nanoplankton, microflagellates, ciliates, and microzooplankton. Because most members of this microbial community do not readily sink from the euphotic zone, excreted labile carbon from the metabolically active microbial community is recycled within the pelagic zone and is potentially available to larger consumers. Another important consequence of the loop is the ability of bacteria to sequester dissolved nitrogen and phosphorus at extremely low concentrations, thereby keeping major limiting nutrients in the epilimnion and minimizing losses to the hypolimnion and sediments. Because bacteria have the ability to outcompete phytoplankton for phosphorus at low concentrations (Currie and Kalff 1984), the potential exists for the development of a large sink of carbon, phosphorus, and other nutrients which are tied up in bacterial biomass. However, because of active predation by flagellates, ciliates, microzooplankton, and some cladocerans, nutrients and DOM are recycled by this active heterotrophic component of the food web, and carbon is returned, albeit inefficiently, to the conventional food chain. Remineralization of nutrients in the new paradigm is now considered more the result of predation and excretion than of direct bacterial decomposition in the pelagic zone (Goldman and Caron 1985).

As alluded to by Azam et al. (1983), the carbon and nutrient flows within the microbial community are tightly coupled, and the dynamic behavior of the loop is basically the result of three processes: *commensalism* (production of DOM by phytoplankton and utilization by bacteria), *competition* (for nutrients between bacteria and phytoplankton, which is influenced by substrate availability), and *predation* (by flagellates, ciliates and microzooplankton, which provides the feedback of nutrients and some DOM).

Because of the potential of these processes for facilitating the growth and production of all components of the pelagic community, one can appreciate why the processes of the microbial loop play such an important role in aquatic systems, especially oligotrophic ones. In such nutrient devoid ecosystems, the retention of DOM and recycling of scarce nutrients in the euphotic zone is crucial for support of ecosystem productivity. However, in more productive meso- and eutrophic lakes and in coastal marine regions, substantial allochthonous inputs of labile organics and available nutrients can markedly affect the outcome of competition between phytoplankton and bacteria and change the relative importance of various pathways of material and energy flow. In these more productive pelagic regions, the overall significance of the microbial loop is diminished and the more direct links (e.g., diatoms-macrozooplankton-fish) of the classic food chain are accentuated. Knowledge of variables that trigger the activity (engagement) of the microbial loop in pelagic communities as a result of seasonal fluxes of temperature, substrates and nutrients along trophic gradients is paramount to our understanding of the role of protists in support of production of pelagic communities.

There is already evidence to suggest that water temperature may strongly influence pathways of carbon flow within microbial components of pelagic communities in both marine (Pomeroy and Diebel 1986) and freshwater ecosystems (Sorokin and Paveljeva 1978; George and Harris 1985). The presence or absence of certain keystone herbivores (e.g., *Daphnia* vs.

Diaptomus) may also influence the pathways of carbon flow and the degree of activity of microbial components in nutrient recycling in the epilimnion of oligotrophic lakes (Stockner and Shortreed 1988; Scavia et al. 1986). Top-down control by planktivorous fish will also influence activities of the microbial loop, as has already been demonstrated by Riemann and Sondergaard (1986) in enclosure experiments in eutrophic Danish lakes.

The succession of bacteria, flagellates and ciliates in microzones of activity surrounding algal cells, zooplankton and particulate organic matter have only recently been studied in freshwater systems (Lehman and Scavia 1984). These microzones may be hot spots of activity, with abundances and productivities that exceed those of surrounding waters by an order of magnitude. New information on these short-lived, microscale processes may further increase the importance of microbial components in our view of freshwater systems.

Conclusions and Recommendations

It is imperative that limnologists begin to incorporate the processes operating in the microbial component of the planktonic food web in their experimental designs and in the interpretations of their results. In future studies requiring separation of heterotrophic from autotrophic components of the plankton using 0.6–1.0 um membranes, it will be important to verify good separation of bacteria from algae (Stockner and Antia 1986) and microflagellates (K. Porter, unpublished data). Direct assessment of particle ingestion through observation and enumeration will be superior to bulk size fractionation experiments, due to the potential inefficiency of separation as well as the new information gained from direct observation of individual species of grazers. Phyto- and zooplankton ecologists must realize that unless extremely rigorous separation procedures are used, their flasks, field containers, or chemostats will contain the metabolically active components of the microbial loop with its associated processes of predation and nutrient regeneration.

It is conceivable that production estimates from lakes have been underestimated owing to the heterotrophic activity of microbial protists, either through respiration, competition, or grazing, during long-term incubations (>4 h). More research is required, but until further results are available, short-term incubations (<1 h) should be adopted as a standard procedure. Inhibitors, tracer particles, fixatives, dilution techniques and size fractionation should be used with prior testing for unpredictable effects in each system studied (Sanders et al. 1985).

Studies of heterotrophic and autotrophic flagellates in fresh water must include experiments to determine the degree of phagotrophy by what were previously considered to be autotrophic phytoplankton. Studies of the seasonal wax and wane of their populations must now consider not only bottom-up control by nutrients, light and temperature, but also top-down control by predators and particulate resource limitation. Investigators studying the dynamics of deep chlorophyll peaks in lakes must recognize the ability of some phytoplankton to supplement their autotrophic nutrition by grazing on picoplankton which often display peak abundance in the vicinity of the thermocline (Bird and Kalff 1986; Pick and Caron 1987). Limnologists need to develop an easy to use method of assessing the strength of microbial interactions in the pelagic, so that the relative rates of nutrient regeneration ascribed to macro- vs microzooplankton can be determined by season and over a spectrum of trophic conditions.

Acknowledgements. We gratefully acknowledge R. Sanders, S. Bennett, A. DeBiase, E. MacIsaac, and K. Shortreed for their insights and the use of their unpublished data. This work was supported in part by NSF grant BSR-84-07928 to KGP. It is contribution No. 35 of the Lake Oglethorpe Limnological Station.

References

Azam, F., T. Fenchel, J. G. Field, J. S. Ray, L. A. Meyer-Reh and F. Thingstad. 1983. The ecological role of water-column microbes in the sea. Mar. Ecol. Prog. Ser. 10:257–263.

Bailey-Watts, A. E., M. E. Bindloss and J. H. Belcher. 1968. Freshwater primary production by a blue-green alga of bacterial size. Nature, London. 220:1344–1345.

Bienfang, P. K. and M. Takahashi. 1983. Ultraplankton growth rates in a subtropical ecosystem. Mar. Biol. 76:213–218. Bird, D. F. and J. Kalff. 1986. Bacterial grazing by planktonic lake algae. Science, Washington, D.C. 231:493–495.

Boraas, M. E., C. C. Remsen and D. D. Seale. 1985. Phagotrophic flagellate populations in Lake Michigan: use of image analysis to determine numbers and size distribution. EOS 66:1299.

Caron, D. A., F. R. Pick and D. R. S. Lean. 1985. Chroococcoid cyanobacteria in Lake Ontario: vertical and seasonal distributions during 1982. J. Phycol. 21:171–175.

Chang, V. T. P. 1980. Zwei neue Synechococcus-Arten aus dem Zurichsee. Schweiz. Z. Hydrol. 42:247–254.

Costella, A. C., K. S. Shortreed and J. G. Stockner. 1979. Phytoplankton fractionation studies in Great Central Lake, British Columbia: a nutrient enriched sockeye salmon (*Oncorhynchus nerka*) nursery lake. Fish. Mar. Serv. Tech. Rep. 800: 27 p.

Craig, S. R. 1984. Productivity of algal picoplankton in a small meromictic lake. Verh. Int. Ver. Limnol. 22:351–354.

Craig, S. R. 1985. Distribution of algal picoplankton in some European freshwaters. Abstr. 2nd Int. Phycol. Congr. Copenhagen, Aug. 1985, 31.

Cronberg, G. and C. Weibull. 1981. *Cyanodictyon imperfectum*, a new chroococcal blue-green alga from Lake Trummen. Sweden Arch. Hydrobiol. Suppl. 60:101–110.

Currie, D. J. and J. Kalff. 1984. A comparison of the abilities of freshwater algae and bacteria to acquire and retain phosphorus. Limnol. Oceanogr. 29:298–310.

Daley, R. J. 1979. Direct epifluorescence enumeration of native aquatic bacteria: uses, limitations, and comparative accuracy. *in*: Native aquatic bacteria: enumeration, activity, and ecology, ed. J. W. Costerton and R. R. Colwell, 29–45. Philadelphia: American Society for Testing and Materials.

Fahnenstiel, G. L., L. Sicko-Goad, D. Scavia and E. F. Stoermer. 1986. Importance of picoplankton in Lake Superior. Can. J. Fish. Aquat. Sci. 43:235–240.

Fenchel, T. 1982a. Ecology of heterotrophic microflagellates. I. Bioenergetics and growth. Mar. Ecol. Prog. Ser. 8: 225–231.

Fenchel, T. 1982b. Ecology of heterotrophic microflagellates. II. Bioenergetics and growth. Mar. Ecol. Prog. Ser. 8: 225–231.

Fenchel, T. 1984. Suspended marine bacteria as a food source. *in*: Flows of energy and material in marine ecosystems—theory and practice, ed. M. R. Fasham, 301–315. New York: Plenum Press.

Fenchel, T. 1986. The ecology of heterotrophic microflagellates. *in*: Advances in microbial ecology, ed. K. C. Marshall, 57–97. New York: Plenum Press.

Fogg, G. E. 1986. Picoplankton. Proc. R. Soc. Lond. B 228: 1–30.

Gates, M. A. 1984. Quantitative importance of ciliates in the planktonic biomass of lake ecosystems. Hydrobiologia 108: 233–238.

George, D. G. and G. P. Harris. 1985. The effect of climate on long-term changes in the crustacean zooplankton biomass of Lake Windermere, U.K. Nature 316(8):536–539.

Gilbert, J. J. and K. G. Bogdan. 1984. Rotifer grazing: *in situ* studies and selectivity rates. *in*: Trophic interactions within aquatic ecosystems, ed. D. G. Meyers and J. R. Strickler. Denver: Westview Press.

Gliwicz, Z. M. 1980. Filtering rates, food size selection, and feeding rates in cladocerans— another aspect of interspecific competition in filter-feeding zooplankton. *in*: Evolution and ecology of zooplankton communities, ed. W. C. Kerfoot, 282–291. Hanover: University Press of New England.

Goldman, J. C. and D. A. Caron. 1985. Experimental studies on an omnivorous microflagellate: implications for grazing and nutrient regeneration in the marine microbial food chain. Deep Sea Res. 32:899–915.

Gude, H., B. Haibel and H. Muller. 1985. Development of planktonic bacterial populations in a water column of Lake Constance (Bodensee-Obersee). Arch. Hydrobiol. 105(1): 59–77.

Hobbie, J. E., O. Holm-Hansen, T. T. Packard, L. R. Pomeroy, R. W. Sheldon, J. P. Thomas and W. J. Wiebe. 1972. A study of the distribution and activity of miocroorganisms in ocean water. Limnol. Oceanogr. 17:544–555.

Johannes, R. E. 1965. Influence of marine protozoa on nutrient regeneration. Limnol. Oceanogr. 10:434–442.

Johnson, P. W. and J. McN. Sieburth. 1982. *In situ* morphology and occurrence of eucaryotic phototrophs of bacterial size in the picoplankton of estuarine and oceanic waters. J. Phycol. 18:318–317.

Lehman, J. T. and D. Scavia. 1984. Measuring the ecological significance of microscale nutrient patches. Limnol. Oceanogr. 29:214–216.

McDonough, R. J., R. W. Saunders, K. G. Porter and D. L. Kirchman. 1986. Depth distribution of bacterial production in a stratified lake with an anoxic hypolimnion. Appl. Environment. Microbiol. 52:922–1000.

Melack, J. M., P. Kilham and T. R. Fisher. 1982. Responses of phytoplankton to experimental fertilization with ammonium and phosphate in an African soda lake. Oecologia 52:321–326.

Meyer, J. L., R. T. Edwards and L. Carlough. 1985. Microbial food web of a Southeastern blackwater river. EOS 66(51): 1334.

Munawar, M. and G. L. Fahnenstiel. 1982. The abundance and significance of ultraplankton and microalgae at an offshore station in Central Lake Superior. Can. Tech. Rep. Fish. Aquat. Sci. 1153:1–13.

Pace, M. L. 1982. Planktonic ciliates: Their distribution, abundance, and relationship to microbial resources in a monomictic lake. Can. J. Fish. Aquat. Sci. 39:1106–1116.

Pace, M. L. and J. D. Orcutt, Jr. 1981. The relative importance of protozoans, rotifers, and crustaceans in a freshwater zooplankton community. Limnol. Oceanogr. 26:822–830.

Pace, M. L., K. G. Porter and Y. S. Feig. 1983. Species and age specific differences in bacterial resource utilization by two co-occurring cladocerans. Ecology 64:1145–1156.

Paerl, H. W. 1977. Ultraphytoplankton biomass and production in some New Zealand lakes. N.Z. J. Mar. Freshwater Res. 11: 297–305.

Peterson, B. J., J. E. Hobbie and J. F. Haney. 1978. *Daphnia* grazing on natural bacteria. Limnol. Oceanogr. 23:1039–1044.

Pick, F. R. and D. A. Caron. 1987. Pico- and nanoplankton biomass in Lake Ontario: Relative contribution of phototrophic and heterotrophic communities. Can. J. Fish. Aquat. Sci. 44: in press.

Platt, T. and W. K. W. Li. 1987. Photosynthetic picoplankton. Can. Bull. Fish. Aquat. Sci. 214:583 pp.

Pomeroy, L. R. 1974. The ocean's food web, a changing paradigm. Bioscience 24:499–504.

Pomeroy, L. R. 1984. Significance of microorganisms in carbon and energy flow in marine ecosystems. *in*: Current perspectives in microbial ecology, ed. M. J. Klug and C. A. Reddy, 405–411. Washington, D.C.: American Society for Microbiology.

Pomeroy, L. R. and D. Diebel. 1986. Temperature regulation of bacterial activity during the spring bloom in Newfoundland coastal waters. Science, Washington, DC. 233:359–361.

Porter, K. G. 1973. Selective grazing and differential digestion of algae by zooplankton. Nature, London. 244: 179–180.

Porter, K. G. 1984. Natural bacteria as food resources for zooplankton. *in*: Current perspectives in microbial ecology, ed. M. J. Klug and C. A. Reddy, 340–345. Washington, D. C.: American Society of Microbiology. Porter, K.G. and Y.S. Feig. 1980. The use of DAPI for identifying and counting aquatic microflora. Limnol. Oceanogr. 25:943–948.

Porter, K. G., M. L. Pace and J. F. Battey. 1979. Ciliate protozoans as links in freshwater planktonic food chains. Nature, London. 277:563–565.

Porter, K. G., Y. Feig and E. Vetter. 1983. Morphology, flow regimes, and filtering rates of *Daphnia, Ceriodaphnia* and *Bosmina* fed natural bacterioplankton. Oecologia 58:156–163.

Porter, K. G., E. B. Sherr, B. F. Sherr, M. Pace and R. W. Sanders. 1985. Protozoa in planktonic food webs. J. Protozool. 32:409–415.

Riemann, B. 1985. Potential influence of fish predation and zooplankton grazing on natural populations of freshwater bacteria. Appl. Environ. Microbiol. 50:187–193.

Riemann, B. and M. Sondergaard. 1986. Regulation of bacterial secondary production in two eutrophic lakes and in experimental enclosures. J. Plankton Res. 8:519–536.

Robbins, E. I., K. G. Porter and K. A. Kaberyan. 1985. Pellet microfossils: Possible evidence for metazoan life in early Proterozoic time. Proc. Nat. Acad. Sci. 82:5809–5813.

Rodhe, W. 1955. Productivity: can plankton production proceed during winter darkness in subarctic lakes? Verh. Int. Ver. Limnol. 12:117–122.

Sanders, R. W., K. G. Porter and R. McDonough. 1985. Bacteriovory by ciliates, microflagellates and mixotrophic algae: factors influencing particle ingestion. EOS. 66: 1314.

Scavia, D., D. A. Laird and G. L. Fahnenstiel. 1986. Production of planktonic bacteria in Lake Michigan. Limnol. Oceanogr. 31:612–626.

Scavia, D. and G. A. Laird. 1987. Bacterioplankton in Lake Michigan: Dynamics, controls, and significance to carbon flux. Limnol. Oceanogr. 32:1017–1032.

Sherr, B. and E. Sherr. 1983. Enumeration of heterotrophic microprotozoa by epifluorescence microscopy. Estuarine Coastal Shelf Sci. 16:1–7.

Sherr, E. B., E. Sherr, R. D. Fallon and S. Y. Newell. 1986. Small, aloricate ciliates as a major component of the marine heterotrophic nanoplankton. Limnol. Oceanogr. 31:177–183.

Shortreed, K. S. and J. G. Stockner. 1986. Trophic status of 19 subarctic lakes in the Yukon Territories. Can. J. Fish. Aquat. Sci. 43:797–805.

Sorokin, Y. I. and E. B. Paveljeva. 1978. On structure and functioning of ecosystems in a salmon lake. Hydrobiologia 57:525–48.

Stemberger, R. S. and J. J. Gilbert. 1985a. Body size, food concentration, and population growth in planktonic rotifers. Ecology 66:1151–1159.

Stemberger, R. S. and J. J. Gilbert. 1985b. Assessment of threshold food levels and population growth in planktonic rotifers. Arch. Hydrobiol. Beih. 21:269–275.

Stenson, J. A. 1985. Interactions between pelagic metazoan and protozoan zooplankton: an experimental study. Verh. Int. Ver. Limnol. 22:3001.

Stockner, J. G. 1987. Lake fertilization: the enrichment cycle and lake sockeye (Oncorhynchus nerka) production, p. 198–215. In H. D. Smith, L. Margolis, and C. C. Wood [ed.] Sockeye salmon (Oncorhynchus nerka) population biology and future management. Can. Spec. Publ. Fish. Aquat. Sci. 96.

Stockner, J. G. 1981. Whole lake fertilization for the enhancement of sockeye salmon (Oncorhynchus nerka) in British Columbia, Canada. Verh. Int. Ver. Limnol. 21: 293–299.

Stockner, J. G. and N. J. Antia. 1986. Algal picoplankton from marine and freshwater ecosystems: a multidisciplinary perspective. Can. J. Fish. Aquat Sci. 43:2472–2503.

Stockner, J. G. and Shortreed, K. S. 1988. Algal picoplankton and contribution to food webs in oligotrophic British Columbia Lakes. Hydrobiologia (in press).

Williamson, C. E. and N. M. Butler. 1986. Predation on rotifers by the suspension-feeding calanoid copepod Diaptomus pallidus. Limnol. Oceanogr. 31:393–402.

Wood, E. J. F. 1965. Marine microbial ecology. New York: Reinhold Publishing Corp.

CHAPTER 6

From Picoplankton to Fish: Complex Interactions in the Great Lakes[1]

Donald Scavia and Gary L. Fahnenstiel

Interpretation of ecological data can be clouded by the influence of unidentified processes and interactions within the system under study. Controversies regarding appropriate methods and interpretation of data are often fostered by this lack of understanding of the ecosystem as a whole. The primary production controversy is an example of this problem. ^{14}C-based estimates of primary production have been claimed to grossly underestimate actual production (Sheldon et al. 1973; Verduin 1975; Tijssen 1979; Schulenberg and Reid 1981; Jenkins 1982). These claims brought the mainstay of primary production estimation under close scrutiny for several years, with much effort centered on the physiological basis for the presumed disagreement (Hobson et al. 1976; Dring and Jewson 1982; Williams et al. 1983; Smith and Platt 1984). It was not until the 1980s that the full impact of the controversy and its relationship to our ideas of carbon flow in the ecosystem became apparent. The controversy was evidently fueled in part by an unperceived complexity of many aquatic ecosystems (Peterson, 1980, Smith et al. 1984). During the initial period of the controversy, conceptualizations of aquatic ecosystems, as exemplified by the extant models (e.g., Steele 1974; Scavia 1979), emphasized traditional food webs that included nano and net phytoplankton, crustacean zooplankton, and fishes; the ^{14}C methodologies were developed consistent with those ecosystem components. It has been discovered more recently, however, that substantial carbon and nutrients flow through the very small organisms (<3 μm) of the so-called microbial food web (Li et al. 1983; Sherr and Sherr 1984). The presence of this microbial loop can have significant influence on food web dynamics and on the interpretation of results from standard ^{14}C experiments. In the relatively long ^{14}C incubations (12-24 hr), designed for the slower growing organisms of the original paradigm, it is likely that much of the ^{14}C is rapidly cycled within the incubation bottle (i.e., fixed, consumed, respired) and the amount of ^{14}C retained on a filter at the end of an incubation may be very different from the amount of

[1]GLERL Contribution No. 564

carbon actually fixed. Thus, it appears that ignorance of the presence and activity of these very small and rapidly turning over organisms was a component of the controversy concerning interpretation of ^{14}C-based production estimates.

In much the same way, our lack of understanding of interactions within an ecosystem can cloud potential cause-effect relationships through the food web. We will describe scenarios for interactions within Lake Michigan indicating that changes at the top of its food web may have observable manifestations four or five trophic levels below. We further suggest that a simple difference in species replacements at one trophic level can force the ecosystem to respond in very different ways.

Great Lakes' Upper Food Web Interactions

The Great Lakes' ecosystem alterations have become classic stories of large lake responses to influence directed intentionally and unintentionally at both the tops and bottoms of their food webs. From as early as the 1800s, the Great Lakes have been manipulated through pollution (Harris and Vollenweider 1982; Beeton 1969) and fishery activities. The classic sequence of fish-related events (Christie 1974) includes lake trout (*Salvelinus namaycush*) overexploitation and infestation by the exotic sea lamprey (*Petromyzon marinus*), leading to catastrophic declines in these top carnivore populations, and subsequent invasion by still another exotic, the planktivorous marine alewife (*Alosa pseudoharengus*). This sequence has apparently not ended. Further alterations have been observed in Lake Michigan and, to varying degrees, in others of the Great Lakes. Tremendously successful sea lamprey control and salmonine stocking programs have produced large populations of salmon and trout (Wells 1985) that support a recreational and commerical fishery valued at more than one billion dollars per year (Talhelm et al. 1979). In Lake Michigan, a combination of increasing alewife predation by these planted salmonines (Stewart et al. 1981) and recent severe winters (Eck and Brown 1985; alewife are particularly susceptible to overwinter stress and die-offs) have forced decreasing alewife abundances between 1970 and 1982 and persistent low abundance levels since 1983. While the alewife decrease was followed by dramatic increases in other fish species both inshore and offshore, the particular fish species and plankton responses in those regions have differed considerably. Nearshore, alewife has been replaced by yellow perch (*Perca flavescens*) and offshore, primarily by bloaters (*Coregonus hoyi*). These replacement species have been extremely successful (Wells 1985; Jude and Tesar 1985).

The burgeoning nearshore yellow perch populations and consequent increased zooplanktivory have produced a 10-fold decrease in zooplankton abundance and biomass. While total zooplankton biomass has been decreasing in the inshore region since 1978 and has remained at very low levels since 1982, there have been no significant changes in species composition or mean body size (Evans 1986). The population continues to be dominated by the small cladoceran, *Bosmina longirostris*. Phytoplankton total cell abundance in inshore regions of southeastern Lake Michigan (off Bridgeman, Michigan) have increased during the same decade (Ayers and Feldt 1983; Bowers et al. 1986). This change, like the zooplankton decrease, has been gradual and may have been a result of reduced zooplankton grazing pressure, although available data suggest that the phytoplankton increase may have preceded the zooplankton decrease by 1 yr. Limited seasonal coverage reported for zooplankton (July-August; Evans 1986) and horizontal phytoplankton patchiness in the relatively variable nearshore environment (Bowers et al. 1986) may have masked the actual long-term trends.

While nearshore plankton have changed rather gradually with little apparent alteration in species composition, the offshore food web has responded differently. Bloater abundance has increased as alewife abundance has decreased, but the former species is predominantly hypolimnetic and the alewife is epi-metalimnetic. This transfer of zooplanktivory from surface waters to the hypolimnion released most zooplankton from the voracious and size selective predation imposed by alewife. The zooplankton community responded with size and species shifts culminating in a switch from summer dominance by calanoid copepods (*Diaptomus* spp.) to dominance by *Daphnia* (Scavia et al. 1986a; Evans and Jude 1986). The replacement was abrupt and occurred between 1982-1983 with overall species dominance in 1983 by *Daphnia pulicaria*, a new species for Lake Michigan. These changes were followed by continued species rearrangements within the daphnid complex among *D. pulicaria* and the two extant species, *D. retrocurva* and *D. galeata mendotae* (Evans and Jude 1985; Scavia and Fahnenstiel 1987). Low alewife abundance and this new daphnid complex persisted in surface waters through at least 1986 (D. Scavia unpublished data).

A number of other characteristics of the offshore lower food web have also changed and remained at this new condition as well. As has occurred in other lakes (e.g., Shapiro and Wright 1984; Lampert 1978; Edmondson and Litt 1982), water clarity increased as *Daphnia* became a major component of Lake Michigan zooplankton. Summer Secchi depths (averaged for the period between the onset of thermal stratification and the beginning of calcite precipitation "whitings") increased from ca. 7.0 ± 1.3 m during the 1975-1982 period to 11.1 ± 1.4 m during 1983-1986, with reports as deep as 17 m in recent years (Fig. 6.1). This large Secchi depth increase was accompanied by increases in other measures of water clarity, including in situ transmissivity vertical profiles and depth attenuation of photosynthetically active radiation (Scavia et al. 1986a; Fahnenstiel and Scavia 1987a). Larger subthermocline phytoplankton communities also accompanied increased epilimnetic transparency (Fahnenstiel and Scavia 1987c).

While total phytoplankton carbon concentration changed only slightly

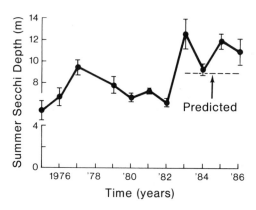

FIGURE 6.1. Offshore Secchi depth, averaged (± S.D.) between the beginning of thermal stratification and the onset of calcite "whiting." Predicted change in Secchi depth is based on the observed chlorophyll concentration change and a model developed for the Great Lakes.

over this time period, there were significant shifts in species composition. The summer epilimnetic phytoplankton of the 1970s and early 1980s typically included 20–40% (by weight) filamentous and colonial blue-green and green algae; those species were much less abundant after 1982 (Fahnenstiel and Scavia 1987a). As had zooplankton species composition and water clarity, the phytoplankton species composition changed abruptly. The species composition determined in 1982 was very much like that reported from the mid-1970s, but that determined after 1982 was overwhelmingly dominated by small (3–30 μm) phytoflagellates. These shifts to new phytoplankton and zooplankton species arrangements in offshore surface waters have been linked to decreased alewife abundance and reduced zooplanktivory both empirically, through process-oriented field studies (Scavia and Fahnenstiel 1987; Fahnenstiel and Scavia 1987a,b), and theoretically, through model simulations of Lake Michigan ecosystem transition during the 1975–1985 period (Scavia et al. 1988).

The Microbial Loop

While ecosystem responses to alterations at the tops of Great Lakes food webs are receiving considerable attention (Kitchell et al. 1988), the focus has generally been limited to the traditional food web (i.e., crustacean zooplankton and nanno- and net phytoplankton). However, much of the production and community metabolism in the Upper Great Lakes occurs within the microbial food web. Nearly half of the primary productivity in the Upper Great Lakes is found in organisms passing a 3-μm screen (Fahnenstiel et al. 1986; G. Fahnenstiel unpublished data) and at times bacterial production can rival total primary production in Lake Michigan (Scavia et al. 1986b; Scavia and Laird 1987).

A critical question regarding the significance of the microbial food web lies in the source vs sink question; i.e., is significant carbon passing from the microbial loop to the traditional food web? Most arguments concerning this question center on determining the number and efficiencies of food web links between the base of the microbial loop (autotrophic and heterotrophic picoplankton) and the location of its interface with the traditional web. The location of this interface within the traditional food web can be altered significantly by higher trophic level interactions. For example, Riemann (1985) demonstrated with large-scale enclosures in eutrophic Frederiksborg Slotsso that the presence or absence of planktivorous fish can alter the role of crustacean zooplankton in transferring bacterial heterotrophic carbon production to the upper food web. In those experiments, the major response to fish addition was reduction in overall zooplankton biomass (as we have seen in the nearshore region of Lake Michigan) and increase in heterotrophic flagellate abundance, with flux of bacterial carbon predominantly to the flagellates. In enclosures without fish, crustacean biomass (main *Daphnia* species) was higher, flagellate abundance was lower, and the major flux of bacterial carbon was to *Daphnia*. Thus, increased zooplanktivory hindered transfer of bacterial carbon to the traditional food web. In those experiments, *Daphnia* apparently dominated the lake and all enclosures regardless of fish abundance. Results from those manipulations are similar to the long-term Lake Michigan nearshore trend, except the dominant Lake Michigan nearshore cladoceran is *Bosmina longirostris* (Evans 1986) instead of *Daphnia*. One could hypothesize that transfer from the microbial food web has become less efficient in Lake Michigan's nearshore.

The situation is likely different offshore; rather than a change in total zooplankton biomass, there was a shift from dominance by *Diaptomus* to dominance by *Daphnia*. The significance of this shift is based on the fact that, while *Daphnia* has the ability to ingest bacteria (0.2-1.2 μm) and micrograzers (2-4 μm) (e.g., Peterson et al. 1978; Borsheim and Olsen 1984; Porter 1984; Riemann and Bosselmann 1984), previously dominant diaptomids cannot ingest bacteria and ingest the small micrograzers at only 20-30% of the efficiency of food of their preferred size (Vanderploeg and Scavia 1979; Vanderploeg et al. 1984).

The importance of this link lies in the hypothesis that the cladocerans are not only capable of capturing bacteria and their predators, but can graze enough to transfer significant quantities of this microbial carbon to the traditional food web. We have estimated rates of micrograzer bactivory (presumably by heterotrophic flagellates and ciliates) for offshore Lake Michigan (Scavia et al. 1986b; Scavia and Laird 1987). While those rates were generally similar to bacterial production rates, the micrograzers alone can balance bacterial production only in spring and late fall (Fig. 6.2). Because bacterial abundance does not change dramatically during summer, a loss other than that due to micrograzers is implied. This imbalance between bacterial production rates and micrograzer predation increases when crustacean abundance increases. In 1984, the zooplankton was quickly dominated by *Daphnia* (Fig. 6.2). If we assume that these crustaceans can make up most of the difference between bacteria production and micrograzer loss measurements, then *Daphnia* may be cropping as much as 70% of Lake Michigan summer epilimnetic bacteria production directly. Riemann (1985) estimated 80-90% transfer of bacterial production to *Daphnia* in his enclosures that lacked fishes.

The presence of such a link between the microbial and traditional food webs does not alone ensure its importance. One measure of its importance comes from comparison of flux of this heterotrophically produced carbon to flux from autotrophic production (Ducklow et al. 1986, 1987; Sherr et al. 1987). While estimates of bacterial carbon production are equivocal (Scavia et al. 1986), our current estimates are that as much as 20 ug C l^{-1} d^{-1} of epilimnetic bacteria production is not grazed by microheterotrophs. Even if a relatively small portion of this is cropped by *Daphnia*, it is significant to *Daphnia* when compared to our current estimate of summer carbon flux from autotrophic nanno and net plankton (8.7 $\pm$ 2.5 ug C l^{-1} d^{-1}, Scavia and Fahnenstiel 1987).

An additional link between the microbial and traditional food webs may also be provided by *Daphnia* predation on the micrograzers (i.e., the heterotrophic or mixotrophic flagellates). These 2–5 μm cells are within the size range of edible prey for *Daphnia*. Estimating carbon production through the heterotrophic flagellate loop is even more speculative; however, based on our estimates of flagellate bactivory (Scavia and Laird 1987) and an assumed flagellate growth efficiency of 37% (Sherr and Sherr 1984); 3.1 ug C l^{-1} d^{-1} would also be available to *Daphnia* via that route. It appears then that there can be significant carbon flux from heterotrophic

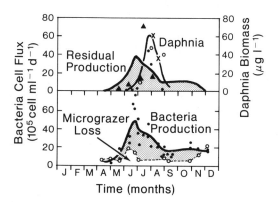

FIGURE 6.2. Bacterial production, micrograzer induced loss, and *Daphnia* abundance in epilimnetic offshore waters. Residual production is the calculated difference between bacterial production and loss due to grazing by microzooplankton.

bacteria production to the traditional food web via *Daphnia* predation on bactivores, as well as on bacteria directly.

The microbial loop is not limited to heterotrophy. Autotrophic pico-plankton are also an important part of the microbial food web in the Great Lakes, and these autotrophs have two distinct differences from hetero-trophic bacteria. First, they are primary producers, and thus their activity is one level higher than that of the heterotrophic bacteria. As much as 50% of total primary production appears to be entering the food web via these picoplankton (Caron et al. 1985; Fahnenstiel et al. 1986; G. Fah-nenstiel, unpublished data). Second, unlike heterotrophic bacteria, au-totrophic picoplankton abundance in Lakes Ontario, Michigan, and Huron exhibit pronounced seasonality. Autotrophic picoplankton abundance in-creases by at least a factor of five from winter-spring to late summer-early fall (Caron et al. 1985; G. Fahnenstiel, unpublished data). The im-portant contribution of autotrophic picoplankton, in terms of both pro-duction and biomass, occurs at a time when the metazoan community is dominated by organisms that can effectively graze them. Thus, autotrophic picoplankton production may also be transferred directly up the food web from the microbial loop at a time when autotrophic picoplankton pro-duction is greatest.

Large bodied *Daphnia*, as opposed to the smaller and more selective *Diaptomus* species, may evoke an efficient link to the microbial loop by shunting newly generated particulate organic carbon (both autotrophic and heterotrophic) directly from several microbial trophic levels (i.e., heter-otrophic bacteria, autotrophic cyanobacteria, heterotrophic and mixo-trophic microflagellates) to the traditional food web in the form of their own body mass, a form that is highly susceptible to vertebrate predation. Complex interactions play an important role in structuring the aforemen-tioned scenario. The differences in feeding strategies of *Daphnia* and *Diaptomus* suggest that the potential significance of the microbial loop as a source of carbon for the traditional food web is influenced strongly by which crustacean species dominates. As discussed above, the summer species arrangements in Lake Michigan are driven to a large extent by predation and interactions higher in the traditional food web.

The Detrital Loop

Changes in vertebrate zooplanktivory may have influenced both water clarity and the interface between the microbial and traditional food webs in another way. Detritus dynamics may have been altered by the recent increase in *Daphnia* summer abundance. The increase in Secchi disc transparency that coincided with increased *Daphnia* abundance was ac-tually greater than that predicted from summer epilimnion chlorophyll concentration changes. An empirical model for the Great Lakes relating summer epilimnion chlorophyll concentration to Secchi depth (Chapra and

Dobson 1981), predicts a Secchi depth increase of 2.8 m for the observed chlorophyll concentration change (2.52 to 1.22 ug l^{-1}, calculated from the difference between 1980-1982 and 1983-1986 summer means). Observed Secchi depth change over that same time period was actually 4.5 m (Fig. 6.1). We suggest that the greater than expected increase in water clarity could have been due to a decrease in nonalgal carbon concentration; *Daphnia* may ingest more detritus than do *Diaptomus*.

The notion of detritus as a large, slowly recycling pool of organic material may also be overstated, especially for systems dominated by nonselective feeders like *Daphnia*. We have examined POC and algal carbon balances for Lake Michigan's summer epilimnion (Scavia and Fahnenstiel 1987) and concluded that, while over 95% of POC leaving the epilimnion was nonliving material, at least 80% of that material was newly generated in the epilimnion. Thus, while *Daphnia* may ingest more detritus than the previously dominant calanoids, they also appear to *produce* large quantities as well, suggesting a more rapidly turning over POC pool. Preliminary calculations, based on information reported by Scavia and Fahnenstiel (1987), and assuming detritus is cleared by *Daphnia* at the same rate as algae, yield a detritus turnover time of only 5 days. If detritus is of recent origin, rather than old and more refractory, then the potential for significant microbial activity associated with it may also increase.

Based on these scenarios, we suggest that as *Daphnia* became dominant in the Lake Michigan summer epilimnion, they may have decreased the standing crop of detritus, yielding clearer water and increased detrital turnover times, and potentially increasing detritus' utility as a microbial substrate.

Conclusion

The preceding arguments are summarized in Fig. 6.3, which implies that changes at the top of the food web have resulted in changes at the trophic levels below. Solid lines represent trends that are supported by direct observations; dashed lines are for those properties less well studied or simply hypothesized. Because the alewife population decline has been tied to predation by the stocked salmonines (Stewart et al. 1981) super-imposed upon poor recruitment and overwinter survival (Eck and Brown 1985), it appears that actions at the very top of Lake Michigan's food web, sea lamprey control and carnivore stocking, have decreased alewife abundance. The complexity of the Lake Michigan food web interactions is best illustrated by comparison of nearshore and offshore plankton responses to the same stimulus (i.e., decrease in alewife abundance). Offshore, total plankton biomass did not change substantially, but both phytoplankton and zooplankton species composition did, and the water became clearer. Nearshore, zooplankton abundance decreased but the population remained dominated by small-sized *Bosmina*. Gradual increases

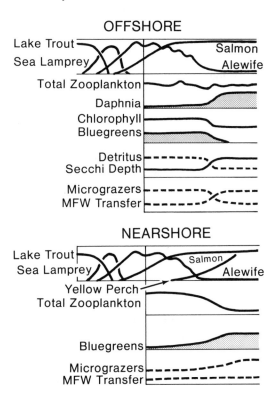

FIGURE 6.3. Schematic representation of observed (solid lines) and hypothesized (dashed lines) trends of plankton abundance in the offshore and nearshore regions of southern Lake Michigan.

in phytoplankton abundance also occurred during the same period. From a food web perspective, the main difference between the two regions was in the fish species replacements for alewife. Offshore, deepwater bloaters and sculpins appear to be the replacements and their primarily hypolimnetic existence *decreased* midsummer epilimnetic zooplanktivory. However, because nearshore alewife have been replaced by substantial populations of yellow perch, zooplanktivory in that region *increased*. The nature of these replacements is likely linked to, among other factors, thermal and morphometric requirements of the replacing species. The results illustrate different directions the ecosystem may take, even within the same lake, given the complex set of biotic and physical interactions and differing environmental constraints.

Because our analyses of these long-term trends were necessarily drawn from a variety of sources representing different levels of spatial and temporal coverage, Fig. 6.3 is best considered as a model of our hypotheses regarding the interactions among the trophic levels from fish to picoplankton in Lake Michigan. Many of the cause-effect implications were based

on extrapolation of laboratory-derived relationships and inferential calculations from field data. The apparent importance of the inferred food
web interactions clearly demand continued efforts toward more process
related field measurements of predator-prey interactions in aquatic ecosystems.

References

Ayers, J. C. and L. E. Feldt. 1983. Benton Harbor power plant imnological studies.
Part XXXI. Phytoplankton of the seasonal surveys of 1980, 1981, and April 1982
and further pre- vs. post-operational comparisons at Cook Nuclear Plant. Special
Report No. 44, Great Lakes Research Division, The University of Michigan,
Ann Arbor.

Beeton, A. M. 1969. Changes in the environment and biota of the Great Lakes.
In: Eutrophication: Causes, consequences, and correctives. Washington, D.C.:
National Academy of Sciences.

Borsheim, K. Y. and Y. Olsen. 1984. Grazing activities by *Daphnia pulex* on
natural populations of bacteria and algae. Verh. Int. Verein. Limnol. 22:644–
648.

Bowers, J. A., R. Rossmann, J. Barres, and W. Y. B. Chang. 1986. Phytoplankton
populations of southeastern Lake Michigan 1974–1982. *in*: Impact of the Donald
C. Cook Nuclear Plant, Publ. 22, Great Lakes Research Division, ed. R. Ross-
man. Ann Arbor: The University of Michigan.

Caron, D. A., F. R. Pick, and D. R. S. Lean. 1985. Chroococcoid cyanobacteria
in Lake Ontario: Vertical and seasonal distribution during 1982. J. Phycol.
21:171–175.

Chapra, S. C. and H. F. N. Dodson. 1981. Quantification of the lake trophic
typologies of Naumann (surface quality) and Thienemann (oxygen) with special
reference to the Great Lakes. J. Great Lakes Res. 7:182–193.

Christie, W. J. 1974. Changes in fish species composition of the Great Lakes. J.
Fish. Res. Board Can. 31:827–854.

Dring, M. J. and D. H. Jewson. 1982. What does ^{14}C uptake by phytoplankton
really measure? A theoretical modelling approach. Proc. R. Soc. Lond. B.
214:351–368.

Ducklow, H. W., D. A. Purdie, P. LeB. Williams, and J. M. Davies. 1986. Bacterioplankton: A sink for carbon in a coastal marine plankton community. Science
232:865–867.

Ducklow, H. W., D. A. Purdie, LeB. P. Williams, and J. M. Davies. 1987. Response to Sherr et al. 1987. Science 235:88–89.

Eck, G. V., and E. H. Brown, Jr. 1985. Lake Michigan's capacity to support lake
trout and other salmonines: An estimate based on the status of prey populations
in the 1970s. Can. J. Fish. Aquat. Sci. 42:449–454.

Edmondson, W. T., and A. H. Litt. 1982. *Daphnia* in Lake Washington. Limnol.
Oceanogr. 27:272–293.

Evans, M. S. 1986. Recent major declines in zooplankton populations in the inshore
region of Lake Michigan: Probable causes and implications. Can. J. Fish. Aquat.
Sci. 43:154–159.

Evans, M. S. and D. J. Jude. 1986. Recent shifts in Daphnia community structure in southeastern Lake Michigan: A comparison of the inshore and offshore. Limnol. Oceanogr. 31:56–67.

Fahnenstiel, G. L. and D. Scavia. 1987a. Dynamics of Lake Michigan phytoplankton: Recent changes in surface and deep communities. Can. J. Fish. Aquat. Sci. 44:449–508.

Fahnenstiel, G. L. and D. Scavia. 1987b. Dynamics of Lake Michigan phytoplankton: Primary production and phytoplankton growth. Can. J. Fish. Aquat. Sci. 44:509–514.

Fahnenstiel, G. L. and D. Scavia. 1987c. Dynamics of Lake Michigan phytoplankton: The deep chlorophyll layer. J. Great Lakes Res. 13:285–295.

Fahnenstiel, G. L., L. Sicko-Goad, D. Scavia, and E. F. Stoermer. 1986. Importance of picoplankton in Lake Superior. Can. J. Fish. Aquat. Sci. 43:235–240.

Harris, G. P. and R. A. Vollenweider. 1982. Paleolimnological evidence of early eutrophication in Lake Erie. Can. J. Fish. Aquat. Sci. 39:618–626.

Hobson, L. A., W. J. Morris, amd K. T. Pirquet. 1976. Theoretical and experimental analysis of the ^{14}C technique and its use in studies of primary production. J. Fish. Res. Bd. Canada 33:1715–1721.

Jenkins, W. J. 1982. Oxygen utilization rates in the North Atlantic subtropical gyre and primary production in oligotrophic systems. Nature 300:246–248.

Jude, D. J. and F. T. Tesar. 1985. Recent changes in the forage fish of Lake Michigan. Can. J. Fish. Aquat. Sci. 42: 1154–1157.

Kitchell, J. F., M. S. Evans, D. Scavia, and L. B. Crowder. 1988. Food web regulation of watter quality in Lake Michigan. J. Great Lakes Res. (in press).

Lampert, W. 1978. Climatic connections and planktonic interactions as factors controlling the regular succession of spring algal bloom and extremely clear water in Lake Constance. Verh. Internat. Verein. Limnol. 20:969.

Li, W. K. W., D. V. Subba Rao, W. G. Harrison, J. C. Smith, J. J. Cullen, B. Irwin, and T. Platt. 1983. Autotrophic picoplankton in the tropical ocean. Science 219:292–295.

Peterson, B. J. 1980. Aquatic primary production and the ^{14}C-CO_2 method: a history of the productivity problem. Annu. Rev. Ecol. Syst. 11:359–385.

Peterson, B. J., J. E. Hobbie, and J. F. Haney. 1978. *Daphnia* grazing on natural bacterial. Limnol. Oceangr. 23: 1039–1044.

Porter, K. G. 1984. Natural bacteria as food resources for zooplankton. *in*: Current Perspectives in Microbial Ecology, ed. M. J. Klug and C. A. Reddy, 340–345. Washington, D.C.: American Society of Microbiology.

Riemann, B. 1985. Potential influence of fish predation and zooplankton grazing on natural populations of freshwater bacteria. Appl. Environ. Microbiol. 50:187–193.

Riemann, B. and S. Bosselmann. 1984. *Daphnia* grazing on natural populations of bacteria. Verh. Internat. Verein. Limnol. 22:795–799.

Scavia, D. 1979. Examination of phosphorus cycling and control of phytoplankton dynamics in Lake Ontario with an ecological model. J. Fish. Res. Bd. Canada. 36:1336–1346.

Scavia, D. and G. L. Fahnenstiel. 1987. Dynamics of Lake Michigan phytoplankton: Mechanisms controlling epilimnetic communities. J. Great Lakes Res. 13:103–120.

Scavia, D., G. A. Lang, and J. F. Kitchell. 1988. Dynamics of Lake Michigan plankton: A model evaluation of nutrient loading, competition, and predation. Can. J. Fish. Aquat. Sci. 45:165–177.

Scavia, D. and G. A. Laird. 1987. Bacterioplankton in Lake Michigan: Dynamics, controls, and significance to carbon flux. Limnol. Oceanogr. 32:1017–1033.

Scavia, D., G. L. Fahnenstiel, M. S. Evans, D. Jude, and J. T. Lehman. 1986a. Influence of salmonine predation and weather on long-term water quality trends in Lake Michigan. Can. J. Fish. Aquat. Sci. 43:435–443.

Scavia, D., G. A. Laird, and G. L. Fahnenstiel. 1986b. Production of planktonic bacteria in Lake Michigan. Limnol. Oceanogr. 31:612–626.

Schulenberg, E. and J. C. Reid. 1981. The Pacific shallow oxygen maximum, deep chlorophyll maximum, and primary productivity reconsidered. Deep-Sea Res. 28:901–919.

Shapiro, J. and D. I. Wright. 1984. Lake restoration by biomanipulation. Freshwat. Biol. 14:371–383.

Sheldon, R. W., W. H. Sutcliff, and A. Prakash. 1973. The production particles in the surface waters of the ocean with particular reference to the Sargasso Sea. Limnol. Oceanogr. 18:719–733.

Sherr, E. B., B. F. Sherr, and L. J. Albright. 1987. Bacteria: Link or Sink? Science 235:88.

Sherr, B. F. and E. B. Sherr. 1984. Role of heterotrophic protozoa in carbon and energy flow in aquatic systems. *in*: Current perspectives in microbial ecology, ed. M. J. Klug and C. A. Reddy, 412–423. Washington, D.C.: American Society of Microbiology.

Smith, R. E. H., R. J. Geider, and T. Platt. 1984. Microplankton productivity in the oligotrophic ocean. Nature 323:252–254.

Smith, R. E. H. and T. Platt. 1984. Carbon exchange and ^{14}C tracer methods in a nitrogen-limited diatom, *Thalassiosira pseudonana*. Mar. Ecol. Prog. Ser. 16:75–87.

Steele, J. H. 1974. The structure of marine ecosystems. Cambridge: Harvard Univ. Press.

Stewart, D. J., J. F. Kitchell, and L. B. Crowder. 1981. Forage fishes and their salmonid predators in Lake Michigan. Transactions of Amer. Fish. Soc. 110:751–783.

Talhelm, D. R., R. C. Bishop, K. W. Cox, N. W. Smith, D. N. Steinnes, and A. L. W. Tuomi. 1979. Current estimates of Great Lakes fisheries values: 1979 status report, Great Lakes Fishery Commission, Ann Arbor, Michigan.

Tijssen, S. B. 1979. Diurnal oxygen rhythm and primary production in the mixed layer of the Atlantic Ocean at 20° N. Neth. J. Sea Res. 13:79–84.

Vanderploeg, H. A. and D. Scavia. 1979. Calculation and use of selective feeding coefficients: Zooplankton grazing. Ecological Modelling 7:135–150.

Vanderploeg, H. A., D. Scavia, and J. R. Liebig. 1984. Feeding rate of *Diaptomus sicilis* and its relation to selectivity and effective food concentration in algal mixtures and in Lake Michigan. J. Plankton Res. 6:919–941.

Verduin, J. 1975. Photosynthetic rates in Lake Superior. Verh. Int. Verein Limnol. 14:134–139.

Wells, L. 1985. Changes in Lake Michigan's prey fish population with increasing salmonid abundance, 1962–1984. *in*: Papers presented for the Council of Lake Committees plenary session of Great Lakes predator-prey issues, March 20,

1985. ed. R. L. Eschenroeder, 13-26. Great Lakes Fishery Commission Special Publication 85-3, Ann Arbor, MI.

Williams, P. Leb., K. R. Heinemann, J. Marra, and D. A. Purdie. 1983. Comparison of ^{14}C and O_2 measurements of phytoplankton production in oligotroiphic waters. Nature 305:49–50.

Part 3
Multiple Causality and Temporal Pattern in Lake Ecosystems

CHAPTER 7

Temporal Variation in Regulation of Production in a Pelagic Food Web Model

S. M. Bartell, A. L. Brenkert, R. V. O'Neill, and R. H. Gardner

Current ecological thinking offers two alternative explanations of measured structure in phytoplankton assemblages. First, interest has been rekindled in the ability of competition theory (Slobodkin 1961; MacArthur 1969; May 1975) to explain patterns of species replacement in phytoplankton (Dugdale 1967; Tilman 1977; Tilman et al. 1981). Competition theory posits that differential rates of use of essential nutrients, relative to rates of their supply, confer advantages in growth rate to different algal species. Changes in algal community structure that result from competition alter the food sources to consumer populations. Changes in the composition of higher trophic levels may ultimately occur. From this perspective, the food web appears structured from the bottom-up (McQueen et al. 1986) in relation to spatial and temporal variability in light intensity, water temperature, and resource availability.

Second, regulation of phytoplankton community structure (O'Neill 1976; Kitchell et al. 1979) may be top-down (McQueen et al. 1986). Selective feeding by omnivorous fishes (O'Brien et al. 1976; Bartell 1982), planktivorous fishes (Brooks and Dodson 1965; Jansenn 1976; Drenner et al. 1982, 1984; Wright and O'Brien 1984), and zooplankton (Peters and Downing 1984) can cascade through lower trophic levels (Carpenter et al. 1985), ultimately determining the structure of the phytoplankton.

Bottom-up and top-down processes occur simultaneously in a dynamic physicochemical context that also influences plankton community structure (Levasseur et al. 1984). Harris (1978, 1983) demonstrated the importance of physical mixing and chemical dynamics in controlling the structure of algal assemblages. These processes may diminish the potential importance of competitive interactions in structuring the phytoplankton by changing the basis for competition over time periods shorter than those required for competitive interactions to eliminate algal populations (Harris 1983). Consumer organisms can contribute directly to the nutrient environment of competing algal populations by augmenting the supply of phosphorus (Peters and Rigler 1973; Lehman 1980; Olsen and Ostgaard 1985) and nitrogen (Blazka et al. 1982; Ejsmont-Karabin 1984).

Observed seasonality in external nutrient loading and temporal variation in the feeding rates of consumers suggested that the relative importance of bottom-up and top-down control of structure in the phytoplankton is scale-dependent. This structure is determined neither entirely by top-down nor bottom-up processes over time scales routinely used in the measurement of phytoplankton community dynamics. To evaluate these ideas, we examined the implications of differential resource use and predator-prey relations in a seasonal environment, formalized as a pelagic food web model, for temporal changes in the relative importance of bottom-up and top-down structuring of the phytoplankton. Detailed analyses of the model's behavior were used to identify periods when control of production shifted from interspecific interactions among the phytoplankton to cascading trophic interactions.

The food web model (FWM) represented a structural compromise between minimal Lotka-Volterra (L-V) population descriptions and comprehensive, site specific lake ecosystem simulators (Park et al. 1974). The FWM simulated the seasonal production of 10 populations of phytoplankton, 5 populations of herbivorous zooplankton, 3 populations of planktivorous fish, and a single population of piscivorous fish (Fig. 7.1). The details of the model are presented in the Appendix.

Model Analyses

Simulations were performed to explore the effects of model structure, system inputs, and population growth parameters on pelagic production dynamics. Daily biomass production of model populations was calculated first over an annual cycle. To characterize the long-term behavior of the model (relative to the characteristic time scales of plankton and fish), 20 yr of daily production were also simulated.

Sensitivity analyses were performed to identify the parameters that determined model performance. We defined regulation of production as the sensitivity of biomass production to small variations in each of 103 FWM parameter values. That is, control of production was assessed in proportion to the relative contribution of individual bottom-up and top-down processes (e.g., nutrient use, photosynthesis, consumption, respiration, predation) to the overall growth of model populations.

Model sensitivities were quantified through repeated simulations in which all 103 parameters were varied simultaneously. The nominal values and coefficients of variation of 2% were used to define normal distributions for each parameter. We emphasize that these distributions have been found useful in previous numerical sensitivity analyses of ecological models (Gardner et al. 1981, Bartell et al. 1986); we do not imply that these distributions accurately represent the variation in values of corresponding

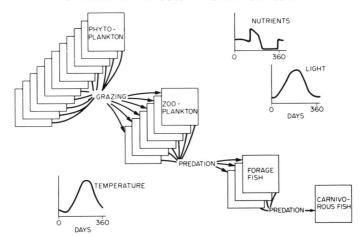

FIGURE 7.1. Schematic illustration of the pelagic food web model (FMW) showing food web structure, trophic relations, and forms of light, temperature, and nutrient functions.

processes in natural systems. For each model prediction of interest, 200 Monte Carlo simulations using parameter values chosen in a stratified random design (Iman and Conover 1980) were performed to quantify the variance in predicted biomass of each of the 19 model populations over yr 2 and 3 (days 240-960) corresponding to the deterministic solution.

Relative partial sums of squares (RPSS) were calculated for each model parameter to determine the amount of variance in model predictions attributable to variation in each parameter (Bartell et al. 1986). The total R^2 of the separate analyses indicated the reliability of the calculated sensitivities and quantified the portion of total variance related linearly to parameter variation. Residual variance $(1—R^2)$ measured the relative importance of nonlinear and interaction effects among parameters in determining production by model populations.

To confirm these model sensitivities, additional sensitivity analyses were performed. We used repeated simulations with parameter values varied as in the previous analyses of days 240-960, but in simulations of a single daily time step at 20 day intervals from days 240-960. Population biomasses and model parameters were reset to the corresponding values of the deterministic solution following the integration of one day's change. Corroboration of the previous sensitivities by these analyses was interpreted as strong evidence of correct identification of parameters that controlled model behavior.

Results

Deterministic Simulations

The FWM produced seasonal changes in the relative abundance of the phytoplankton populations comparable to those commonly measured in pelagic systems (Fig. 7.2). Following ice melt, populations 1-5 progressively dominated the phytoplankton biomass. Under conditions of high light, warm temperatures, and low nutrient availability, populations 6-8 became relatively more abundant. At any given time, several populations contributed to the overall phytoplankton production, although some populations (e.g., 9 and 10) were typically rare.

Deterministic simulations of 20 y of daily production demonstrated the eventual establishment of a 2-yr cycle for all 19 model populations following an initial 3-4 yr period of variable biomass transients (Fig. 7.3). The existence of the cycle was independent of the initial biomass values of the populations, although initial biomass influenced the duration of the population transients prior to the establishment of the cycle.

In the absence of the piscivore population, a 1-yr cycle became established after 5 simulated yr of production. The 2-yr cycle appeared to result from the model interactions between the growth dynamics of the piscivore population and its indirect influence on internal nutrient dynamics (Kitchell et al. 1975). The piscivores acted as a big slow component (O'Neill et al. 1975; Carpenter this volume) of the system with regard to predator-prey feedback on internal nutrient cycling. Elimination of both the planktivorous fish populations and the piscivore population also resulted in a 1-yr cycle.

One-yr cycles were also produced by directly altering the nutrient inputs to the system. A fivefold increase in the daily nutrient input generated a 1-yr cycle in productivity following about 16 yr of transient behavior. Halving the nutrient input rate established a 1-yr cycle following about 5 yr of transient behavior. Establishment of the 1-yr cycle through direct manipulation of nutrient inputs further suggested that the 2-yr cycle resulted from the indirect effects of predation, especially by the piscivore, on nutrient availability to the phytoplankton. The importance of these interactions in system function has also been demonstrated in pelagic system models derived from assumptions different from those underlying the FWM (Bartell 1978; Carpenter and Kitchell 1984) and in experimental studies (Bergquist and Carpenter 1986; Sterner 1986).

Sensitivity Analyses

Our analyses demonstrated temporal variation in the relative importance of bottom-up and top-down regulation of phytoplankton diversity as an implication of the ecological interactions formalized in the food web model (Fig. 7.4). Periods of decreasing algal diversity corresponded to the over-

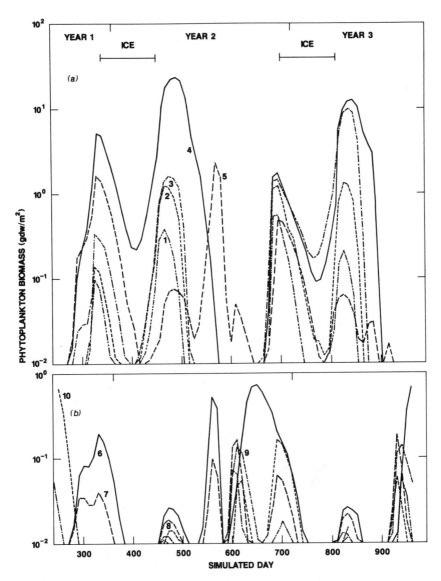

FIGURE 7.2. Seasonal values of biomass of the 10 phytoplankton populations.

riding importance of differences in individual algal population growth parameters in determining the relative abundance of the 10 phytoplankton populations. The importance of competitive interactions highlighted these periods, although in the varying model environment no single population eliminated all of its competitors (Fig. 7.2).

The highest values of phytoplankton diversity occurred when top-down regulation by the consumer populations was most evident (Fig. 7.4). Var-

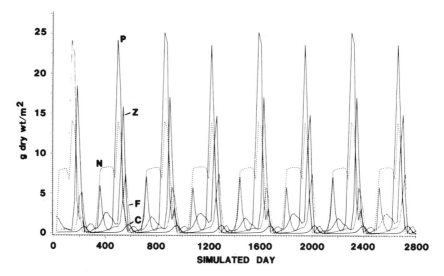

FIGURE 7.3. Summary of nutrient concentration (N) and biomass of phytoplankton (P), zooplankton (Z), planktivorous fish (F), and piscivorous fish (C) for an 8 yr deterministic simulation.

iation in the consumption and feeding parameters of the piscivorous fish population indirectly produced greater changes in phytoplankton diversity during these times than did variation in parameters that defined algal physiology in terms of eq. 1–4 and Table 7.2.

Examination of the sensitivity of a single phytoplankton population to parameter variations underscored the temporal pattern of regulation demonstrated for the entire algal assemblage (Fig. 7.5). Phytoplankton population 4 was selected for detailed analysis because it contributed the larg-

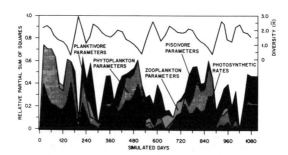

FIGURE 7.4. Relative importance of parameters for different trophic levels in explaining variance in phytoplankton diversity over a 3 yr period. The relative partial sums of squares were calculated from 200 Monte Carlo simulations where parameters were varied by 2% of their mean values.

TABLE 7.1. Values of relative partial squares of sensitive FWM parameters produced by the local sensitivity analysis of the phytoplankton

Simulated day	FWM parameter						
	$P_{max_{1j}}$	To_{1j}	Is_{1j}	k_{1j}	W_{1j}	CM_{2j}	To_{2j}
261	0.169	0.203		0.049		0.084	0.285
281	0.067	0.148				0.066	0.443
301	0.226	0.031		0.128		0.023	0.235
321	0.048	0.053					
341	0.028	0.039	0.023				
361	0.059		0.049				
381	0.112	0.058	0.090		0.036		0.050
401	0.240	0.238	0.180		0.039		0.073
421	0.197	0.205	0.171				
441	0.197	0.205	0.118				
461	0.179	0.080					
481	0.920						
501	0.826						
521	0.353	0.305			0.121		0.072
541	0.047	0.869			0.026		
561	0.047	0.852					
581	0.028	0.411			0.068		0.023
601	0.036						0.034
621	0.172	0.138		0.038		0.083	0.289
641	0.239	0.313	0.042	0.069		0.044	0.230

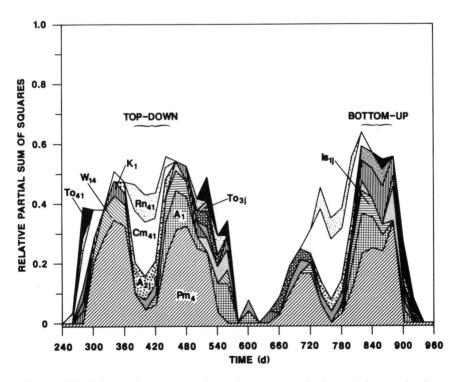

FIGURE 7.5. Relative importance of model parameters in determining production by phytoplankton population 4. Parameters are defined in the Appendix.

est portion of the total annual primary productivity. Population 4 was grazed primarily by zooplankton population 5. Thus, values of susceptibility of algal population 4 to grazing by zooplankton population 5 (W14), assimilation by zooplankton 5 (A14), and the temperature optimum of this zooplankton population (To25) were all important in determining the zooplankton grazing pressure experienced by algal population 4. Because zooplankton population 5 was in turn preyed upon primarily by planktivorous fish population 3, the productivity of algal population 4 was sensitive to variation in the feeding rate of this planktivore population (Cm33). (The first subscript refers to the trophic level, for example, 1 = phytoplankton, 2 = zooplankton, etc., and the second subscript identifies the specific population within the trophic level.)

In addition to the relative sensitivity of production by algal population 4 to seasonal zooplankton grazing and predator-prey interactions throughout the food web, the results demonstrated periods where competitive interactions with other phytoplankton populations largely determined productivity of algal population 4. The productivity of population 4 was sensitive to variation in the half saturation constant, temperature optimum, and light saturation constant of algal population 3. The temperature optimum of algal population 5 was also important in determining productivity by algal population 4.

The 1-day sensitivity analyses largely confirmed the results of the previous analyses and added new information concerning the behavior of the FWM. When parameter variability was permitted to propagate throughout the system for only a single day, parameters of populations in nonadjacent trophic levels failed to rank in importance. The 1-day analysis of total phytoplankton production demonstrated the short-term importance of the phytoplankton and zooplankton growth parameters in the FWM. The broad pattern of temporal switching between the importance of algal parameters that controlled production through differential growth in relation to changing light, temperature, and nutrient conditions and the control of production exerted through grazing pressure by zooplankton was retained in these analyses. The important difference was that when parameters were varied for only a single day, feedback of higher trophic levels on the production dynamics of zooplankton failed to become established. This results from the daily time step of the model and the direct mathematical coupling of adjacent trophic levels only.

The RPSS for the total phytoplankton analysis (Table 7.1) indicated temporal changes in the relative importance of algal parameters Pmax1j, To1j, Is1j, and Xk1j (j = 1,10) in controlling short term algal productivity. The zooplankton parameters that were important were susceptibility of the algae to grazing (W2j), zooplankton feeding rate (Cm2j), and the temperature optima for the zooplankton (To2j), jk = 1,5. For the time periods listed, the values of the RPSS for each parameter varied by nearly an order of magnitude, thus demonstrating the degrees of influence that model

TABLE 7.2. List of parameters used to define 10 populations of
phytoplankton in light, temperature, and nutrient resource space

Population	Optimal temperature T_o (C)	Light saturation $w_o s$ (ly/d)	Nutrient constant k (g/m²)	Maximum photosynthesis P_m (1/d)
1	10	100	0.19	1.6
2	12	114	0.17	1.8
3	14	128	0.15	2.0
4	16	142	0.13	2.1
5	18	156	0.11	2.0
6	20	170	0.09	1.9
7	22	184	0.07	1.8
8	24	198	0.05	1.6
9	26	212	0.03	1.5
10	28	226	0.01	1.3

parameters can exert on biomass production during different periods of
simulation.

The time varying influence of algal population parameters on production
dynamics of the piscivore population was also demonstrated (Fig. 7.6).
Thus, bottom-up regulation of food web production was measurable at
the highest trophic level, just as piscivore growth parameters importantly
influenced phytoplankton production at certain times. While phytoplankton
parameters were occasionally important in indirectly determining piscivore
production, parameters of the zooplankton and planktivorous fishes explained more of the variance in piscivore production. Depending on the
time period, the relative importance of zooplankton and planktivore parameters was reversed. For certain periods of the simulations, the indirect
effects of the other populations were more important than piscivore physiology in determining piscivore production dynamics. The temporally
shifting patterns of bottom-up and top-down regulation were also quantified
for the zooplankton and planktivorous fishes in the FWM. These patterns,

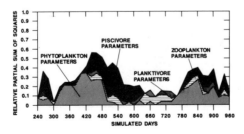

FIGURE 7.6. Temporal changes in the relative importance of model parameters in
explaining variance in production of piscivore biomass for days 240–960.

while not presented here, further emphasized the indirect interconnections between the phytoplankton and the piscivore populations.

Bottom-up and top-down regulation of phytoplankton production was interrelated through the separate influences of algae and consumers on the concentration of available nutrients (Carpenter and Kitchell 1984; Bartell et al. 1988). Bottom-up regulation used nutrients to build algal biomass which then passed to higher trophic levels. The consumer populations altered nutrient availability as a consequence of their growth dynamics. During periods of ice cover or low water temperatures and corresponding low feeding rates of consumers, modeled nutrient concentrations were primarily influenced by rates of phytoplankton growth (Fig. 7.7). Light and temperature constraints on algal growth during these periods, coupled with an abundance of nutrients, decreased the importance of differences in nutrient affinities among the 10 algal populations in determining nutrient concentrations. Between days 500-680, consumer production was more important in indirectly determining nutrient concentrations.

Discussion

Bottom-Up or Top-Down Regulation

The potential for shifts between bottom-up and top-down control of phytoplankton production was implicit in the structure of eq. 1, where growth was represented as an integration of rates of population gains and losses. In a constant environment without consumer populations, production would be determined entirely by light, temperature, and nutrient constraints (eq. 2) imposed on the maximum growth rates of the populations, further mediated by competitive interactions (Tilman 1977). Indeed, our

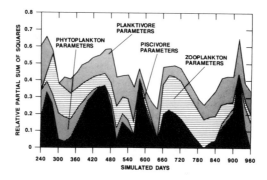

FIGURE 7.7. Temporal sensitivities of dissolved nutrient concentration to variation in model parameters for days 240–960.

analyses showed that phytoplankton parameters were most important in determining phytoplankton biomass during periods of rapid growth (Fig. 7.5). An associated cost of rapid accumulation of algal biomass was an accelerated loss of gross primary production to consumers (eq. 7). Eventually, grazing pressures exceeded the production capacity, which was constrained by reduced nutrient supplies, and phytoplankton biomass declined (Fig. 7.3). Model analyses demonstrated increased sensitivity of phytoplankton production to variation in consumer population parameters during these periods of declining algal biomass.

The evidence for bottom-up and top-down structuring argues that both processes are important, yet neither process by itself comprehensively explains measured patterns of diversity in the phytoplankton. Our sensitivity analyses demonstrated that temporal shifts in relative importance of bottom-up and top-down control of phytoplankton assemblages are a necessary consequence of food web components interacting on a daily time scale in a seasonally varying environment. This result should be general for any system, terrestrial or aquatic, that resembles the FWM in its basic food-web structure and rate structure.

To ask whether competition or predation is more important in regulating system structure ignores an implicit (and important) assumption of scale. Explanations of phytoplankton community structure framed in an either-or context are simplistic. The model sensitivities indicate that it is theoretically possible to design a single experiment which will show either top-down control, bottom-up control, or inconclusive results, depending on the distribution of biomasses of producer and consumer populations and the underlying rate structure of the system at the time the experiment is performed.

Identification of circumstances where competition or predation processes dominate control, and specification of the conditions where control shifts primarily from one direction to the other are logical steps toward development of a coherent theory concerning the structure and dynamics of the phytoplankton (Harris 1986, Legendre and Demers 1984). Our initial attempts to identify the conditions leading to shifts between alternative patterns of regulation in relation to flux along individual pathways in the FWM were unsuccessful. We speculated that, as a prelude to a shift in the pattern of sensitivities, the relative flux along individual pathways in the system would change substantially. Therefore, we calculated the daily flux of mass along all system pathways and enumerated all terms in the equations for each population. Drastic changes in the patterns of flux or values of factors regulating the growth processes were not observed in relation to the sudden shifts between bottom-up and top-down regulation (Fig. 7.5). Indirect feedbacks involving several connected model pathways likely preordained the abrupt changes in the pattern of sensitivities. Our results suggest that experimental manipulations can shift patterns of control (Carpenter et al. 1987), but that the shifts will not be significantly correlated with increased or decreased flux along any single dominant path through the food web.

Constraints on Production

Analyses of food web characteristics and assumptions embodied in the model lend additional evidence for the hierarchical control of phytoplankton production and community dynamics (c.f. Levasseur et al. 1984). The FWM sensitivities identified a set of nested constraints that regulated production in the pelagic food web. The seasonal pattern and magnitude of light intensities and associated water temperatures ultimately determined the potential for total annual food web production. In separate simulations, scalar increases in light intensity and water temperatures exerted greater effects on production dynamics than did similar variability in available nutrients or values of individual model parameters.

Within the broad constraints delineated by light and temperature regimes, available nutrients exerted more fine scaled (short-term) control of the annual production dynamics of the modeled phytoplankton assemblage. Even finer detail concerning the temporal variation in production by specific populations resulted from the relationships among the particular parameter values assigned to each model population.

Further elaboration of the relative importance of bottom-up and top-down control of production dynamics in aquatic systems may result from refinements of our approach. While the sensitivity analyses support hierarchical control of phytoplankton production dynamics, these analyses pertain to the behavior of a single model when all parameters were varied by the same small amount (i.e., 2% CVs). These analyses focus on the implications of model structure and the sensitivity of the deterministic solution to the nominal parameter values. A more rigorous evaluation of the relative importance of bottom-up versus top-down control requires empirically based estimates of variance for model parameters and more comprehensive aquatic system models (Carpenter; this volume; Harris and Griffiths 1987). Together with manipulative experiments performed in natural systems (Bender et al. 1984; Carpenter et al. 1987), these refinements may further our understanding of bottom-up and top-down control and facilitate the development of a coherent theory of dynamic limnology and oceanography.

Acknowledgement. This research was sponsored jointly by the U.S. Environmental Protection Agency under Interagency Agreement No. DW899306009-01-0, and the National Science Foundation under Interagency Agreement BSR-8315185 with the U.S. Department of Energy under Contract No. DE-AC05-84OR21400 with Martin Marietta Energy Systems, Inc. Publication No. 3093, Environmental Sciences Division, ORNL.

Appendix

Model Description

The model populations at each trophic level were differentiated by parameter values that determined their growth rate in relation to light, temperature, and available nutrient (Tables 7.2 and 7.3). Competitive and trophic interactions represented by constant coefficients (the a_{ij}'s) in the L-V models were, in the FWM, time varying and modeled explicitly in terms of available light, nutrient, and prey. In the FWM, the intrinsic growth rates characteristic of the L-V models were disaggregated into nutrient utilization, photosynthesis, respiration, consumption, and mortality. The model produced seasonal patterns of growth and succession comparable to those commonly reported for northern dimictic lakes, although the FWM was not derived for any particular lake.

Phytoplankton

The model simulated the daily biomass (g dry mass/m²) production of 10 functionally defined populations of phytoplankton. The biomass of each phytoplankton population (B_i) changed differentially in relation to rates of photosynthesis (P_i), an aggregate respiration-mortality term (M_i), and zooplankton grazing (G_i) (O'Neill and Giddings 1979):

$$dB_i/dt = B_i (P_i - M_i - G_i.) \qquad [1]$$

Our objective was to define algal populations that differentially responded to daily changes in surface light intensity (I), water temperature (T), and nutrient availability (N). This was accomplished by formulating population-specific rates of photosynthesis as

$$P_i = (Pm_i f(N) g(I) - R_i) h(T) \qquad [2]$$

where Pm_i was the maximum photosynthetic rate for population i, which was modified as a function of nutrients (f), light (g), and temperature (h). To emphasize competition for resources among these populations, a single aggregate loss term (sinking + respiration + mortality), R_i, of 0.2 d⁻¹ was applied equally to all ten algal populations.

Differential resource utilization and optimal growth conditions were formalized by defining nonlinear functions for f, g, and h. Each of the 10 populations differed according to the parameter values that specified these functions (Table 7.2). The dependence of photosynthesis on available nutrient, f, was

$$f(N) = N / (k + N), \qquad [3]$$

where k was the nutrient concentration (g/m²) which reduced maximum photosynthesis by one-half. This equation was neutral to the implications of cell quotas and nutrient pools internal to phytoplankton (Lehman et

TABLE 7.3. List of parameter values used to define populations of zooplankton, and planktivorous and piscivorous fish

Population	Optimal temperature (C)	Maximum consumption (1/d)	Maximum respiration (1/d)	Susceptibility to predation (unitless)
Zooplankton				
1	12	0.50	0.090	0.450
2	16	0.55	0.080	0.425
3	20	0.60	0.070	0.400
4	24	0.65	0.050	0.375
5	28	0.70	0.015	0.350
Planktivorous fish				
1	14	0.23	0.100	0.575
2	20	0.25	0.080	0.600
3	26	0.27	0.070	0.625
Piscivorous fish				
1	20	0.20	0.100	0.600

al. 1975), yet was compatible in structural resolution to other components and processes represented in the overall model (e.g., no internal food reserves were accumulated within zooplankton or fish).

The influence of light intensity on photosynthesis was modeled after Park et al. (1974):

$$g(I) = 0.316/(c+bZ) \, (e^x - e^y) \qquad [4]$$

where $x = Y \exp[-z(c+bZ/z)]$, $Y = -I/(Is_i)$. Implicit in calculation of the parameters in eq. 4 was a daily fractional photoperiod defined as 0.5 which defined a 12-h light:dark cycle. Light availability was influenced by depth (z) and the sum of biomass of all 10 algal populations (Z). The depth of the euphotic zone was chosen as 10 m. Values of $c = 0.2$ and $b = 0.1$ were defined in terms of their specific light saturation intensities, Is_i (Table 7.2).

Each population was assigned an optimal temperature, To_i, for maximum photosynthesis. The functional form of the temperature dependence (Shugart et al. 1974, Kitchell et al. 1974, Titus et al. 1975) increased photosynthesis up to To, then subsequently decreased it to zero as temperature approached an upper lethal value, Tm_i:

$$h(T) = V^X \exp[x(1-V)], \qquad [5]$$

where $V = (Tm_i - T)/(Tm_i - To_i)$, X was a nonlinear scalar of the ln Q_{10} $(Tm_i - To_i)$, and the Q_{10} determined the slope of $h(T)$ for $0 < T < To_i$. Individual populations were differentiated through by single parameter, To_i, after setting Tm_i to 35 C and Q_{10} to 2.0 for all 10 populations (Table 7.2).

Sinusoidal daily changes in light intensities (I) and water temperatures (T) were the primary model inputs. Nutrient (N) supply was constant at

0.01 g/m^2/d, except for spring and fall turnover when the supply was augmented by 7 g/m^2. Available (= dissolved) nutrient was diminished by phytoplankton growth using the ratio of 8:1 (biomass:nutrient) based on the nitrogen composition of phytoplankton (Bloomfield 1975). Equations 1–5, together with changing conditions of I, N, and T, produced an algal community capable of successional behavior in relation to changing conditions of light, temperature, and nutrients (O'Neill and Giddings 1979).

Consumer Populations

Biomass changed in populations of zooplankton, planktivorous fish, and the piscivore as the net result of consumption (C_i), aggregate respiration-mortality (M_i), and (except for the piscivore population) losses to predation (F_i). These processes were summarized by

$$dB_i/dt = B_i (C_i - M_i - F_i). \qquad [6]$$

This equation for consumer production dynamics has proven useful in previous studies of zooplankton and fish (Kitchell et al. 1974, 1977; Smith et al. 1975; Bartell et al. 1986).

The consumer equation followed Park et al. (1974) and represented feeding as a function of the biomass of the predator, B_i, prey biomass, B_j, for n prey populations; preference of the predator for the prey population, w; and the assimilation of the prey by the predator, defined as a constant 0.8 (unitless) in this study:

$$C = Cm_i h(T) B_i 0.8 < [(w B_j)/(B_i + w B_j)] \qquad [7]$$

Cm_i was the physiological maximum rate of consumption for population i and $h(T)$ was eq. 5, the corresponding To representing the optimal temperature for feeding. When prey were abundant, C was determined primarily by the biomass of the predator. Conversely, prey biomass determined values of C when prey biomasses were low or predators were abundant (DeAngelis et al. 1975).

A physiological maximum respiration rate assigned to each population was modified by water temperature using eq. 5. Each of the consumer populations was further distinguished by its particular rate of consumption and susceptibility to predation (Table 7.3).

References

Bartell, S. M. 1978. Size-selective planktivory and phosphorus cycling in pelagic systems. Ph.D. Dissertation, University of Wisconsin, Madison.

Bartell, S. M., J. E. Breck, R. H. Gardner, and A. L. Brenkert. 1986. Individual parameter perturbation and error analysis of fish bioenergetics models. Can. J. Fish. Aquat. Sci. 43:160–168.

Bartell, S. M. 1982. Relationships between prey abundance and size selective predation by bluegill in Lake Wingra. Trans. Amer. Fish. Soc. 111:453–461.

Bartell, S. M., A. L. Brenkert, and S. R. Carpenter. 1988. Parameter uncertainty and the behavior of a size-dependent plankton model. Ecol. Model. 40:85–95.

Bender, E. A., T. J. Case, and M. E. Gilpin. 1984. Perturbation experiments in community ecology: theory and practice. Ecology 65:1–13.

Bergquist, A. M. and S. R. Carpenter. 1986. Limnetic herbivory: effects on phytoplankton populations and primary production. Ecology 67:1351–1360.

Blazka, P., Z. Brandl, and L. Prochazkova. 1982. Oxygen consumption and ammonia and phosphate excretion in pond zooplankton. Limnol. Oceanogr. 27:294–303.

Bloomfield, J. A. 1975. Modeling the dynamics of microbial decomposition and carbon cycling in the pelagic zone of Lake George, New York. Ph.D. dissertation, Rensselaer Polytechnic Institute, Troy, New York.

Brooks, J. L. and S. I. Dodson. 1965. Predation, body size, and the composition of the zooplankton. Science 150:28–35.

Carpenter, S. R. and J. F. Kitchell. 1984. Plankton community structure and limnetic primary production. Am. Nat. 124:159–172.

Carpenter, S. R., J. F. Kitchell, and J. R. Hodgson. 1985. Cascading trophic interactions and lake productivity. BioScience 35: 634–639.

Carpenter, S. R., J. F. Kitchell, J. R. Hodgson, P. A. Cochran, J. J. Elser, M. M. Elser, D. M. Lodge, D. Kretchmer, X. He, and C. N. von Ende. 1987. Regulation of lake primary productivity by food web structure. Ecology 68:1863–1876.

DeAngelis, D. L., R. A. Goldstein, and R. V. O'Neill. 1975. A model for trophic interaction. Ecology 56:881–882.

Drenner, R. W., F. DeNoyelles, Jr., and D. Kettle. 1982. Selective impact of filter-feeding gizzard shad on zooplankton community structure. Limnol. Oceanogr. 29:941–948.

Dugdale, R. C. 1967. Nutrient limitation in the sea: Dynamics, identification and significance. Limnol. Oceanogr. 12:685–695.

Ejsmont-Karabin, J. 1984. Phosphorus and nitrogen excretion by lake zooplankton (rotifers and crustaceans) in relationship to individual body weights of the animals, ambienttemperature, and presence or absence of food. Ekol. Pol. 32:3–42.

Gardner, R. H., R. V. O'Neill, J. B. Mankin, and J. H. Carney. 1981. A comparison of sensitivity analysis and error analysis based on a stream ecosystem model. Ecol. Model. 12:177–194.

Harris, G. P. 1978. Temporal and spatial cycles in phytoplankton ecology: Mechanisms, methods, models, and management. Can. J. Fish. Aquat. Sci. 37:877–900.

Harris, G. P. 1983. Mixed layer physics and phytoplankton populations: studies in equilibrium and non-equilibrium ecology. Prog. Phycol. Res. 2:1–52.

Harris, G. P. 1986. Phytoplankton ecology. Structure, function, and fluctuation. London: Chapman and Hall.

Harris, G. P. and F. B. Griffiths. 1987. On means and variance in aquatic food chains and recruitment to the fisheries.Freshwat. Biol. 17:381–385.

Iman, R. L. and W. J. Conover. 1980. Small sample sensitivity analysis techniques for computer models with an application to risk assessment. Comm. Stat. A9(17):1749–1842.

Jansenn, J. 1976. Feeding modes and prey size selection in the alewife (*Alosa pseudoharengus*). J. Fish. Res. Board Can. 33:197–219.

Kitchell, J. F., J. F. Koonce, R. V. O'Neill, H. H. Shugart, J. J. Magnuson, and R. S. Booth. 1974. Model of fish biomass dynamics. Trans. Amer. Fish. Soc. 103:786–798.

Kitchell, J. F., D. J. Stewart, and D. Weininger.1977.Applications of a bioenergetics model to yellow perch (*Perca flavescens*) and walleye (*Stizostedion vitreum vitreum*). J. Fish. Res. Board Can. 34:1922–1935.

Kitchell, J. F., J. F. Koonce, and P. S. Tennis. 1975. Phosphorus flux through fishes. Verh. Internat. Verein. Limnol. 19:2478–2484.

Kitchell, J. F., R. V. O'Neill, D. Webb, G. Gallepp, S. M. Bartell, J. F. Koonce, and B. S. Ausmus. 1979. Consumer regulation of nutrient cycling. BioScience 29: 28–34.

Legendre, L. and S. Demers. 1984. Towards dynamic biological oceanography and limnology. Can. J. Fish. Aquat. Sci. 41:2–19.

Lehman, J. T. 1980. Nutrient recycling as an interface between algae and grazers in freshwater communities. *in*: Evolution and ecology of zooplankton communities, ed. W. C. Kerfoot, 251–263. Hanover: University Press of New England.

Lehman, J. T., D. B. Botkin, and G. E. Likens. 1975. The assumptions and rationales of a computer model of phytoplankton population dynamics. Limnol. Oceanogr. 20:343–364.

Levasseur, M., J-C. Therriault, and L. Legendre. 1984.Hierarchical control of phytoplankton succession by physical factors. Mar. Ecol. Prog. Ser. 19:211-222.

MacArthur, R. H. 1969. Species packing, or what competition minimizes. Proc. Nat. Acad. Sci. 64:1369–1375.

May, R. M. 1975. Some notes on estimating the competition matrix, alpha. Ecology 56:737–741.

McQueen, D. J., J. R. Post, and E. L. Mills. 1986. Trophic relationships in freshwater pelagic ecosystems. Can. J. Fish. Aquat. Sci. 43:1571–1581.

O'Brien, W. J., N. A. Slade, and G. L. Vinyard. 1976. Apparent size as the determinant of prey selection by bluegill sunfish (*Lepomis macrochirus*). Ecology 57:1304–1310.

Olsen, Y. and K. Ostgaard. 1985. Estimating release rates of phosphorus from zooplankton: Model and experimental verification. Limnol. Oceanogr. 30:844–852.

O'Neill, R. V. 1976. Ecosystem persistence and heterotrophic regulation. Ecology 57:1244–1253.

O'Neill, R. V., W. F. Harris, B. S. Ausmus, and D. E. Riechle. 1975. A theoretical basis for ecosystem analysis with particular reference to element cycling. *in*: Mineralcycling in Southeastern Ecosystems, ed. F. G. Howell,J. B. Gentry, and M. H. Smith, 28–40. ERDA/DOE CONF-740513.

O'Neill, R. V. and J. M. Giddings. 1979. Population interactions and ecosystem function: phytoplankton competition and community production. *in*: Systems Analysis of Ecosystems, Statistical Ecology Series, Volume 9, ed. G. S. Innis and R. V. O'Neill, 103–123. Fairland, Maryland: International Co-operative Publishing House.

Park, R. A. et al.1974. A generalized model for simulating lake ecosystems. Simulation 23:33–50.

Peters, R. H. and J. A. Downing. 1984. Empirical analysis of zooplankton filtering and feeding rates. Limnol. Oceanogr. 29:763–784.

Peters, R. H. and F. H. Rigler. 1973. Phosphorus release by *Daphnia*. Limnol. Oceanogr. 18:821–829.

Scavia, D., J. A. Bloomfield, J. S. Fisher, J. Nagy, and R. A. Park. 1974. Documentation of CLEANX: A generalized model for simulating the openwater ecosystem of lakes. Simulation 23:51–56.

Shugart, H. H., R. A. Goldstein, R. V. O'Neill, and J. B. Mankin. 1974. TEEM: A terrestrial ecosystem energy model for forests. Oecol. Plant. 9:231–264.

Slobodkin, L. B. 1961. Growth and regulation of animal populations. New York: Holt, Rinehart, and Winston.

Smith, O. L., H. H. Shugart, R. V. O'Neill, R. S. Booth, and D. C. McNaught. 1975. Resource competition and an analytical model of zooplankton feeding on phytoplankton. Am. Nat. 109:571–591.

Sterner, R. W. 1986. Herbivores' direct and indirect effects on algal populations. Science 231:605–607.

Tilman, D. 1977. Resource competition between planktonic algae: an experimental and theoretical approach. Ecology 58:338–348.

Tilman, D., M. Mattson, and S. Langer. 1981. Competition and nutrient kinetics along a temperature gradient: An experimental test of a mechanistic approach to niche theory. Limnol. Oceanogr. 26:1020–1033.

Titus, J., R. A. Goldstein, M. S. Adams, J. B. Mankin, R. V. O'Neill, P. R. Weiler, H. H. Shugart, and R. S. Booth. 1975. A production model for *Myriophyllum spicatum* L. Ecology 56:1129–1138.

Wright, D. I. and W. J. O'Brien. 1984. The development and field test of a tactical model of the planktivorous feeding of white crappie (*Pomoxis annularis*). Ecol. Monogr. 54:65–98.

CHAPTER 8

Transmission of Variance Through Lake Food Webs

Stephen R. Carpenter

Temporal variability is a hallmark of ecological communities. Variable species dynamics imply that certain limnological events will not be predictable (Edmondson 1979) and that the predictability of ecological systems may have inherent limits (Paine 1981). Because temporal variability impedes prediction, it is a source of frustration for ecologists and resource managers. However, there are hopeful signs that certain approaches to ecological variability may yield new insights about community organization and allow more successful management of natural resources. I offer three illustrations of this point.

1. Novel explanations of community organization result from explicit consideration of variability. For example, variable recruitment dynamics are common (Strong 1986). Such variable recruitment reverses trends toward competitive exclusion (Chesson 1986). The resultant variance in population dynamics may be essential for ecosystem energetics; most pelagic food webs would be energetically impossible if predators had to subsist on mean productivities of prey, rather than on peaks in productivity of variable prey populations (Steele 1984).

2. Approaches to resource management under uncertainty are being developed (Walters 1986). Management programs and ecosystem variability interact. For example, fisheries management appears to increase the variability of fishery yields (Regier and Henderson 1973). Such variability cascades through pelagic food webs, increasing variability at all trophic levels and confounding efforts to predict fishery dynamics from limnological variables (Carpenter and Kitchell 1987).

3. The fact that ecological variance tends to be scale dependent (see below) offers the prospect of investigating system structure by studying variance in relation to scale (Allen 1977). Ecologists have made extensive use of the scale dependency of spatial variance in analyzing the pattern of vegetation (Pielou 1977). In aquatic communities, similar spatial analyses have been performed for plankton (Steele 1978) and macrophytes (Carpenter and Titus 1984). Temporal analyses of ecological

processes have received much less attention, though the potential is clearly recognized (Shugart 1978; O'Neill et al. 1986). Some studies of aquatic systems have considered both spatial and temporal scales of variability (Abbott et al. 1982; Kratz et al. 1987).

This chapter explores the relationships between the temporal distribution of variance of primary production in lakes and the relative magnitudes of top-down and bottom-up forces on the phytoplankton. Annual or longer time scales are emphasized. The results complement those of Bartell et al. (this volume), who pursue similar goals in their analysis of a different food web at shorter time scales.

Scale of Variability in Lake Productivity

Variance entails scale. In a stable community at equilibrium, population densities do not change, the variance of successive samples in time is zero, and scale does not matter. However, in a community that is changing through time, the variance of successive samples depends on scale (i.e., the duration of the sampling periods and/or the intervals between them). Examples of stable equilibria are rare in natural communities (Connell and Sousa 1983). The continual fluctuation of natural populations indicates that variance must depend on scale in all community studies.

The temporal variance of primary production is scale dependent (Carpenter and Kitchell 1987). This variance may be apportioned among a hierarchy of factors. The long-term potential productivity of a lake is set by edaphic, morphometric, and hydrologic factors (Schindler 1978; Fee 1979). Meteorological factors that determine ice-out, extent of spring mixing, and summer thermal regime and hydrodynamics cause substantial year to year variability around the long-term potential productivity (Strub et al. 1985, Scavia et al. 1986). Between the spring phytoplankton bloom and fall turnover, herbivory and certain physical phenomena cause further variability in primary production at daily to monthly time scales (Harris 1986; Sommer et al. 1986; Scavia et al. 1986; Carpenter et al. 1987; Mills et al. 1987; Bartell et al. this volume; Mills and Forney this volume). Because a large percentage of ecosystem metabolism occurs during summer (Wetzel 1983), annual primary production can be strongly influenced by herbivory.

The suggestion of Hrbáček et al. (1961) that changes in fish community structure could alter the biomass and community structure of zooplankton and phytoplankton is now clearly confirmed. Several recent papers have summarized two decades of research showing that food web dynamics influence the phytoplankton (DiBernardi 1981; Carpenter et al. 1985; McQueen et al. 1986; Northcote 1988). Effects of fish dynamics have been extended to microbial and detrital aspects of the carbon cycle (Riemann and Sondergaard 1986; Scavia and Fahnenstiel, this volume). At least two whole-lake fish manipulations produced substantial reductions in the phytoplankton (Henrikson et al. 1980; Shapiro and Wright 1984). These studies lacked reference (control) ecosystems and critics pointed out that other changes unrelated to the fish manipulations may have affected the phytoplankton (Post and McQueen 1987). However, whole-lake fish manipulations incorporating a reference ecosystem also showed massive changes in phytoplankton biomass and productivity (Carpenter et al. 1987) and changed nutrient limitation of algae between nitrogen and phosphorus (Elser et al. 1988). The maximum variance in phytoplankton induced by food web change is roughly equal to that induced by extreme fluctuations in spring mixing (Scavia et al. 1986, Carpenter et al. 1987). Understanding the interactions of such biological and meteorological factors remains a major challenge.

Transmission of Variance

Environmental factors, such as weather, nutrient supply, flushing rate, and mixing events affect phytoplankton both directly and indirectly (Fig. 8.1). These factors have both predictable (e.g., seasonal) and stochastic components. Direct effects of weather on nutrient supply, mixing regime, and other physical and chemical factors generally act within a year to alter phytoplankton density and productivity (Harris 1980). Weather also has indirect effects on phytoplankton through its effect on recruitment of fishes (Carpenter and Kitchell 1987). These indirect effects may be highly de-

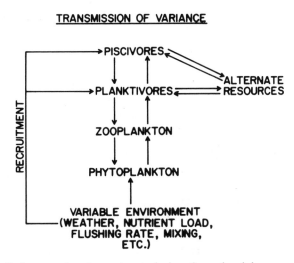

FIGURE 8.1. Pathways of variance transmission through a lake ecosystem. Variability in the lake's environment causes variability in phytoplankton production and in the recruitment of fishes.

pendent on food web structure. Recruitment depends on population size of the parent stock and predation as well as weather (Pitcher and Hart 1982). Recruitment is notoriously variable, and models to predict it have high rates of failure (Sissenwine 1984). Variability in recruitment may have long-term effects on phytoplankton, because effects of a large cohort of piscivores on lower trophic levels persist for the life of the cohort (Carpenter et al. 1985).

A Model of Variance Transmission

To explore the hypothesized difference in time scale between indirect food web effects and the direct effects of weather on phytoplankton, I constructed a stochastic model of a lake food chain (Fig. 8.1). The model was calibrated using literature data reviewed by Carpenter and Kitchell (1987) and data from three lakes used in food web manipulations during 1984-1986 (Carpenter et al. 1987; unpublished data of S. Carpenter, J. Kitchell, J. Hodgson, and associates). Fish life histories and diets simulate those of minnows and largemouth bass in these lakes (Hodgson and Kitchell 1987 and unpublished data). Because the goal of the model is to analyze the structure of interannual variance, it uses an annual time step. Time series generated by the model were subjected to spectral analysis and cross-spectral analysis (Chatfield 1980).

Primary production P varies about the mean productivity P^* at minimum zooplankton biomass ($Z = 1$ g m^{-2}) because of climatic variability and effects of grazing:

$$P = P^* \{(1 + V_p) + [c_1 (1 - Z)]\}. \qquad [1]$$

Meteorological variability in production is a normally-distributed random variable V_p with mean zero and standard deviation 0.2, consistent with available data (Carpenter and Kitchell 1987). The empirical parameter c_1 determines the decrease in productivity for zooplankton biomasses above 1 g m^{-2}.

Deviation from the mean zooplankton biomass in the absence of predation, Z^*, results from meteorological fluctuations in phytoplankton production and planktivory:

$$Z = Z^* + [c_2 (1 + V_p)] - p'F_1 - P'F_2 \qquad [2]$$

The empirical parameter c_2 converts meteorological variability in primary production to variability in zooplankton biomass. The vectors p and P contain age-specific zooplanktivory coefficients for each planktivore. Planktivores include a short-lived fish (age vector F_1) that is planktivorous throughout its 3 yr life cycle, and the juvenile year classes of a longer-lived fish (age vector F_2). Adults of the longer-lived fish are principally piscivorous, but their diets include some zooplankton. Z cannot be lower than 1 g m⁻² in the model. Such buffering from effects of extreme plank-tivory could result from a compositional shift to small zooplankters which are invulnerable to fish predation.

Each fish population is projected from year to year via a matrix A of survivorships (on the subdiagonal) and per capita age specific rates of recruit production (on the top row). Other elements of A are zero. Equations have the same form

$$F_t = A F_{t-1} \qquad [3]$$

but different parameter values for the two fish populations. The survi-vorships of the small fishes decline linearly as piscivore biomass increases. Small fishes are assumed to have a littoral refuge from predation which will shelter a minimum population of 1000 adult fish ha⁻¹. In both fish populations, recruitment per adult fish declines as density of adult fish increases. Such declines could result from cannibalism or other density dependent factors. Each age specific per capita recruitment rate a_{ij} depends on the total adult population T, the carrying capacity for adults K, the age-specific maximum recruitment rate r_j, and recruitment variance which is normally distributed with mean zero and standard deviation V_r:

$$a_{ij} = (1 + V_r) \, r_j \, [1 - (T/K)] \qquad [4]$$

When equation 4 is combined with equation 3, total recruitment is a un-imodal function of the total adult population and the distribution of year class sizes is lognormal, consistent with data from many fish populations (Pitcher and Hart 1982, Walters 1986, Carpenter and Kitchell 1987). If V_r is sufficiently large, recruitment appears to be density-independent at in-termediate levels of the adult population, and density-dependent only when the adult population is very small or very large (cf. Strong 1986). Published V_r values range approximately from 0.25 to 1.5 (Carpenter and Kitchell 1987). Except where noted otherwise, a V_r of 0.5 was used in these sim-ulations.

The population of small fish is semelparous, with reproduction and death occurring in yr 3. The population of large fish is iteroparous. Except where noted otherwise, its life history is modeled after largemouth bass in north-ern Wisconsin. First reproduction occurs at yr 4, and maximum longevity is 10 yr. Year to year survivorship increases with age, and r_j increases with age j after 4 yr.

Botsford (1979) studied the stability properties of a very similar model of a fish population. In Botsford's model, unlike this model, recruitment was deterministic and a certain fraction of the adult stock was harvested each year. In all other respects, Botsford's model is similar to the piscivore component of my model. The critical parameters for stability of Botsford's model were the maximum potential fecundity per adult and the age at first reproduction. Depending on the values of these parameters, the model exhibited asymptotic stability, damped oscillations, stable oscillations, or chaotic behavior (Botsford 1979). The goal of this paper is to examine variance induced by stochastic environmental factors, not variance that results from oscillatory or chaotic behavior of the deterministic part of the model. Therefore, survivorships and r_j values were selected to produce a stable equilibrium point in the absence of environmental variability. This stable equilibrium point provided the initial condition for all stochastic simulations, unless otherwise noted.

Model Results

Time courses generated by the deterministic version of the model have constant primary production, zooplankton biomass, and fish populations.

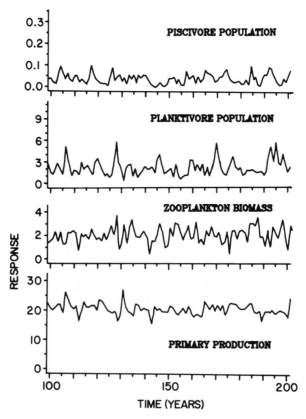

FIGURE 8.2. Example of a time course produced by the model. Years 100–200 of a 512-yr simulation are shown for the piscivore population (i.e., adults of the large, iteroparous fish; thousands of individuals ha^{-1}), planktivore population (juveniles of the large fish plus the population of small, semelparous fish; thousands of individuals ha^{-1}), zooplankton biomass (g dry mass m^{-2}), and primary production during summer stratification (g C m^{-2}).

Stochastic simulations (Fig. 8.2) exhibit the variability known from long-term records of fish and plankton (Soutar and Isaacs 1974; Edmondson and Litt 1982). Casual examination of time courses calculated by the model reveals little pattern.

Spectral analysis of the time course of primary production reveals the underlying variance structure by calculating variance (spectral density) as a function of scale (period in years of each cyclic variance component). When the model is run deterministically, with no added variance, primary production is constant and the variance is therefore zero at all scales. When the variance of fish recruitment is held at zero, but variance is added as the direct effect of environmental variability on the phytoplankton, variance is about the same at all scales (Fig. 8.3A). Such variance is known as white noise because no scale is dominant, by analogy to white light (which is a mixture of all wavelengths of the visible spectrum). When the variability of fish recruitment V_r is set at 0.5, a peak of spectral density appears at scales of 7-10 yr (Fig. 8.3B). When V_r is raised to 1.0, this peak of spectral density intensifies and a harmonic peak appears at a scale of 14-20 yr (Fig. 8.3C).

The approximate correspondence of the scale of the dominant peak of variance of primary production and the life span of the top predator is a consistent feature of this model (Table 8.1). When the piscivorous fish population is set to zero, so the short-lived, semelparous planktivore is the top predator, the variance peak of primary production appears at 3

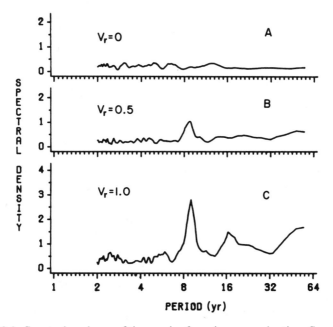

FIGURE 8.3. Spectral analyses of time series for primary production. Spectral density, or variance, is plotted vs period in years in all panels. *A*, Variability of piscivore recruitment $V_r = 0$; *B*, $V_r = 0.5$; *C*, $V_r = 1.0$.

TABLE 8.1. Effects of selected life history characteristics of the top predator on the dominant scale of variance in the primary production time series. Age at first reproduction, lifespan, and the scale (period) of the variance peak of primary production are in years. Relationships to age of fecundity and year to year survivorship are constant or increase or decrease linearly, as indicated

Age at first reproduction	Life span	Age dependent functions for adults		Scale of primary Production
		Fecundity	Survivorship	
3	3	n/a	n/a	3
2	5	increase	increase	4–5
4	8	increase	increase	6–8
4	10	increase	increase	8–10
4	16	increase	increase	11–16
4	10	constant	increase	8–10
4	10	increase	constant	12–22[1]

[1]Piscivore population exhibited long-term oscillations in deterministic run of model.

yr. When the life span of the piscivore was set at 5, 8, 10 or 16 yr, the scale of the variance peak showed corresponding increases and occurred at a scale somewhat shorter than the piscivore life span. This pattern suggests that the functions for fecundity vs age and survivorship vs age may determine the exact position of the variance peak. However, when fecundity vs age was made constant, no appreciable change occurred in the variance peak. When survivorship beyond the age of first reproduction was made constant, the variance peak broadened and shifted to longer periods. This piscivore life history could not be stabilized in deterministic runs of the model, so the variance peak may result from deterministic oscillations of the fish population rather than variance in recruitment.

Some Implications

Time courses of primary production consistently exhibit large variance components at scales near the life span of the dominant predator (Table 8.1). Variance in primary production at the scale of the dominant predator's life span is directly proportional to the variance of the predator's recruitment (Fig. 8.3). These results are an ecosystem-level manifestation of the storage effect (Warner and Chesson 1985). When a fish life history is iteroparous and adults survive well, one large year class per life cycle can sustain the population, even in the face of severe competition and a fluctuating environment (Warner and Chesson 1985). The effects of that large year class on the food web and lake ecosystem will be felt for the life span of the fishes (Carpenter et al. 1985). At certain times of the year, the fishes' effect on primary production will be pronounced (Bartell et al. this volume).

I predict that paleolimnological records of indicators of primary production will contain variance components at time scales near the life spans of piscivores. These scales will be distinct from those of other sources of variance in the paleolimnological record. In chromatic terms, the white noise of climatic forcing and the red (i.e., long wavelength) variance of long-term trends in climate and watershed biogeochemistry lend a pink tint to the paleolimnological spectrum. Most sediment core records will not be fine grained enough to detect the very short wavelength variance of seasonal change, but many will contain the intermediate wavelength variance of piscivore effects. An example of fish effects on algal pigment deposition at a time scale of about 10 yr is known from Lake Michigan (Kitchell and Carpenter 1987).

Scale and Comparative Limnology

The success of regression analyses in studies of the effects of nutrient loading on lakes (Peters 1986) suggests that a similar approach may be useful in studies of food web processes (Pace 1984; McQueen et al. 1986).However, even deterministic models (Oksanen et al. 1981; Persson et al. this volume) suggest highly variable relationships between biomasses of different trophic levels. My results further suggest that investigators' choices of sampling scales will influence the outcome of regression analyses. The warnings of Box et al. (1978, pp. 487-498) concerning regression analysis of dynamic systems with feedback (such as predator-prey systems) are especially pertinent, as scale effects are prominent in such cases. Box et al. recommend that feedback systems be disconnected or dynamic systems be allowed to reach equilibrium before collecting data for regression analyses. The impracticalities of implementing these recommendations in field studies are obvious.

Model results illustrate some of the difficulties of applying regression to food web processes. Cospectral analysis examines the joint relationships of paired time series, and can reveal much about the structure of dynamic systems (Chatfield 1980). Of the several spectra used in cospectral analysis, the coherency spectrum is most germane to regression analysis in ecology. The squared coherency is identical to r^2, the square of the product-moment correlation coefficient (Chatfield 1980).

The coherency spectrum of the piscivore and primary producer time series illustrates the difficulty in analyzing food web dynamics by regression if the incorrect time scale is chosen (Fig. 8.4). The correlation of the piscivore population with primary production is low and quite variable at

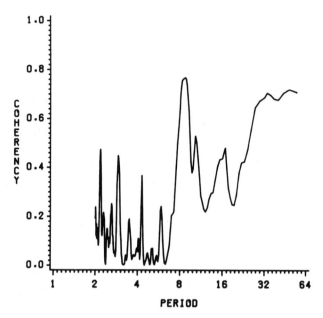

FIGURE 8.4. Coherency squared vs period in years for the piscivore and primary production time series. Coherency squared equals the square of the product–moment correlation coefficient.

scales shorter than the piscivore life span. The correlation is high in the vicinity of the piscivore life span (10 yr), then decreases at scales of 12-30 yr, and becomes high again for scales longer than about 30 yr. Of course, primary production and the piscivore are linked in the model (Fig. 8.1). A type II error would occur if the investigator concluded otherwise on the basis of a low correlation. Low correlations are especially likely at time scales less than 6 yr. Few ecological data bases are long enough to avoid this scale problem. An analogous scale dependent relationship of zooplankton and phytoplankton time series was seen by Carpenter and Kitchell (1987), who found positive correlations at certain scales, and negative correlations at other scales. It is not surprising that regression analyses using data at relatively short scales have found only weak correlations between fish and phytoplankton (McQueen et al. 1986). Weak correlations are expected, even if the causal links are strong. The scale dependencies of Fig. 8.4 help explain why fisheries biologists have found that "the predictive value of empirical models beyond the date of publication is usually unknown" and "post-publication failure of empirical models is not infrequent" (Sissenwine 1984, p. 70).

A great challenge in the analysis of food web processes is to distinguish the variance due to unpredictable environmental fluctuations from that due to temporal fluctuations in largely deterministic ecological processes. Scale dependency makes this problem especially difficult.

Variability and Lake Management

The potentially large effects of fishes on phytoplankton prompted the suggestion that biomanipulation of fisheries could provide a cost effective means of controlling algae (Shapiro 1980, Shapiro and Wright 1984). The biomanipulation concept has been presented to lake managers as a simple series of reciprocal interactions among successive trophic levels; e.g., increased piscivory causes decreased planktivory, increased herbivory, and decreased algal biomass (Shapiro et al. 1982). One critic emphasizes this simplicity and regards biomanipulation as an equilibrium concept that is untenable because of the disequilibrial nature of plankton communities (Harris 1986, pp. 252-258).

Biomanipulation need not entail equilibrium, and biomanipulation methods should anticipate the temporal variability and time scale effects hypothesized by Carpenter et al. (1985) and explored in this paper. Evidence for such temporal variability is growing. For example, in one whole-lake experiment, rapid habitat shifts by young fish caused abrupt fluctuations in primary production at a time scale of about one month, in contrast to the multiyear effects observed in a companion experiment during which fish behavior was more stable (Carpenter et al. 1987 and unpublished data). Such surprising responses are important clues for basic limnologists attempting to develop a theory of food web interactions in lakes, but would evoke disappointment in a management context. Neglect of temporal variability may cause biomanipulation attempts to fail, and lead to premature abandonment of a promising management technique.

The available evidence from limnology and fisheries biology shows that the dynamics of lake ecosystems are quite variable at certain time scales. These time scales include those at which society often asks aquatic ecologists for predictions. However, within the context of any management problem, it is possible to assign probabilities to the different ecosystem responses that derive from specifiable but unpredictable events (Walters 1986). For example, the transparency of Lake Michigan during summer depends on the current planktivore stock and whether or not the southern basin froze over the preceding winter (Scavia et al. 1986). One could use long-term weather records, fishery records, hatchery projections, and so forth to estimate the expected probability distributions of planktivore stocks and ice cover, and hence transparency, for Lake Michigan in the future. Such probabilistic approaches can provide a rational basis for management that does not create false expectations about the certainty of the consequences of management decisions (Walters 1986). Basic aquatic ecology has much to contribute to the development of such models.

Research Strategies for Variable Systems

Progress on both basic and applied problems occurs faster when system-wide manipulations are employed (Walters 1986). To illustrate this point, a forecasting equation was applied to the model of Figure 8.1. The forecasting equation uses the current piscivore stock in three age classes (YOY = young of the year, J = juveniles, and M = mature piscivores) to predict primary production (P) the next year.

$$P_{t+1} = b_o + b_1 YOY_t + b_2 J_t + b_3 M_t + E. \qquad [5]$$

The parameters b_i are estimated by standard regression techniques using the data available up to yr $t+1$. E represents the error of the forecast of P. In the next year of the simulation, observed primary production is compared with the prediction, the forecasting equation is updated to include this new information, and the subsequent year's primary production is forecast. As more years are simulated, the standard deviation of the forecast will tend to decline. The problem is to decrease the standard deviation of the forecast as rapidly as possible.

Uncertainty, measured by the standard deviation of the forecast of primary production, was calculated during two 30 yr runs of the model (Fig. 8.5). In both cases, 6 yr of data were accumulated prior to the first forecast. In the observation run, the system was not disturbed and uncertainty declined gradually. In the experiment run, juvenile and adult piscivore densities were reduced 90% in year 7. Uncertainty declined abruptly between yr 7-9, and remained low for the duration of the simulation.

Uncertainty declines rapidly following large-scale manipulations because such experiments reveal a large portion of the system's potential behavior in a relatively short time. In nonmanipulative studies, researchers must wait for natural variability to reveal the potential range of system behavior. Natural experiments of sufficient severity to produce rapid declines in uncertainty may be quite rare. Walters (1986) argues that uncertainty can actually grow during monitoring programs, and that occasional major manipulations are essential to reduce uncertainty in studies of complex systems.

Weak manipulations (i.e., those that change the independent variables within the range normally experienced by the system) do not reduce uncertainty, and may increase it (Walters 1986).For example, studies employing narrow ranges of the independent variables have prompted controversy about the potential effectiveness of biomanipulation (Pace 1984; Post and McQueen 1987). Both of these studies found that only small differences in the algae could be detected across zooplankton assemblages ranging from bosminids to small daphnids. However, whole-lake studies that have involved changes from small crustaceans to large daphnids (or, in one case, to large herbivorous copepods) have yielded extensive changes in the algae (Henrikson et al. 1980; Shapiro and Wright 1984; Scavia et

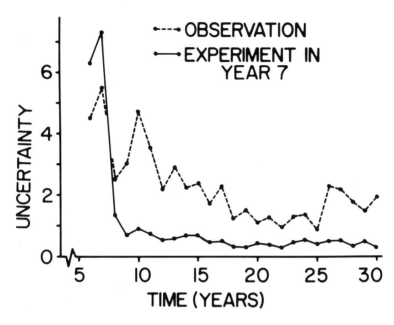

FIGURE 8.5. Uncertainty of the forecasting model vs time (yr) in two simulations. Uncertainty is the standard deviation of the prediction of the subsequent year's primary production from the current year's piscivore stock. In the "observation" simulation, the system is not disturbed. In the "experiment" simulation, piscivore density was reduced 90% in yr 7.

al. 1986; Mills et al. 1987; Carpenter et al. 1987). Thus, the magnitude of the response in the dependent variable (phytoplankton biomass) depends on the range induced in the independent variable (zooplankton size structure).Major manipulations determine the potential responses of complex systems, while effects of minor manipulations may be indistinguishable from background variance.

The history of ecosystem manipulations (Likens 1985), tenets of natural resource management (Walters 1986), and examples presented in this chapter indicate that experimentation on large, heterogeneous systems is a powerful approach to analyzing complex interactions. Yet, simplification of the system to be studied is a common response to the daunting variability of natural systems. Investigations that seek a scale at which variance is acceptably small defer the challenge of abiding (Strong 1986) or embracing (Walters 1986) the variance of natural communities and ecosystems. Experimentation at a smaller scale, perhaps under more controlled conditions, may be the only way to determine if certain mechanisms are possible and how they work. Interpretations of large-scale experiments may be greatly enhanced by results of smaller-scale experiments on similar systems. However, it may be impossible to infer community-wide behavior from studies at only one relatively small scale (O'Neill et al. 1986). In complex systems, compensatory or opposed mechanisms can always be identified. Only properly scaled manipulations can determine which of these mechanisms are important at the community or ecosystem level.

The processes addressed by this chapter (Fig. 8.1) are but one example of a large class of ecological phenomena that involve transmission of variance through networks. For example, any food web that incorporates a variable, strongly interacting species (sensu Paine 1980) should show similar scaling of variance components to the life history of that species. Succession is another example, in which space moves through a network of seral states and the scale of variability is determined by disturbance regimes and the life histories and interactions of colonizing species (West et al. 1981).

We lack a general theoretical perspective on variance transmission through networks. Existing models are either a very particular representation of a single type of system (Fig. 1), deal with only one periodic source of variation (one color), or deal with white noise in which no periodicities dominate. Real ecological noise has multiple colors, imparted by the time delays, life history characteristics, and interspecific interactions that govern the time scales of system response to external sources of variance. Understanding how these sources of variance are transmitted through ecological networks, how they interact with the networks, and how their scales are modified with passage through the networks are unifying problems that cut across all ecological subdisciplines.

Acknowledgements. These ideas owe much to discussions with Jim Kitchell, conducted in coffee shops, canoes, and johnboats throughout the upper midwest. His wise counsel, and the advice of S. M. Bartell, D. M. Lodge, R. P. McIntosh, and L. Persson substantially improved earlier drafts of the manuscript. D. Scavia introduced me to the forecasting literature which led eventually to Figure 8.5, and W. S. Overton suggested that the effects of age-specific fecundity and survivorship on the variance spectra be investigated. This research was supported by NSF grants BSR-83-08918, BSR-86-06271, and BSR-86-18434.

References

Abbott, M. R., T. M. Powell, and P. J. Richerson. 1982. The relationship of environmental variability to the spatial patterns of phytoplankton biomass in Lake Tahoe. J. Plankton Res. 4:927–941.

Allen, T. F. H. 1977. Scale in microscopic algal ecology: A neglected dimension. Phycologia 16:253–258.

Botsford, L. W. 1979. Population cycles caused by inter-age, density-dependent mortality in young fish and crustaceans. *in:* Cyclic phenomena in marine plants and animals, ed. E. Naybr and R. G. Hartnoll, 73–82. New York: Pergamon.

Box, G. E. P., W. G. Hunter, and W. S. Hunter. 1978. Statistics for Experimenters. New York: John Wiley & Sons.

Carpenter, S. R. and J. E. Titus. 1984. Composition and spatial heterogeneity of submersed vegetation in a softwater lake. Vegetatio 57:153–165.

Carpenter, S. R., J. F. Kitchell, and J. R. Hodgson. 1985. Cascading trophic interactions and lake productivity. BioScience 35:634–639.

Carpenter, S. R. and J. F. Kitchell. 1987. The temporal scale of limnetic primary production. Am. Nat. 129:417–433.

Carpenter, S. R., J. F. Kitchell, J. R. Hodgson, P. A. Cochran, J. J. Elser, M. M. Elser, D. M. Lodge, D. Kretchmer, X. He, and C. N. von Ende. 1987. Regulation of lake primary productivity by food web structure. Ecology 68:1863–1876.

Chatfield, C. 1980. The analysis of time series. New York: Halsted.

Chesson, P.L. 1986. Environmental variation and the coexistence of species. in: Community Ecology, ed. J. Diamond and T. J. Case, 240–256. New York: Harper and Row.

Connell, J. H. and W. P. Sousa. 1983. On the evidence needed to judge ecological stability or persistence. Am. Nat. 121: 789–824.

DiBernardi, R. 1981. Biotic interactions in freshwater and effects on community structure. Boll. Zool. 48:353–371.

Edmondson, W. T. 1979. Lake Washington and the predictability of limnological events. Arch. Hydrobiol. Beih. 13:234–241.

Edmondson, W. T. and A. H. Litt. 1982. Daphnia in Lake Washington. Limnol. Oceanogr. 27:272–293.

Elser, J. J., M. M. Elser, N. A. MacKay, and S. R. Carpenter. 1988. Zooplankton-mediated transitions between N and P limited algal growth. Limnol. Oceanogr. 33:1–14.

Fee, E. J. 1979. A relation between lake morphometry and primary productivity and its use in interpreting whole-lake eutrophication experiments. Limnol. Oceanogr. 24:401–416.

Harris, G. P. 1980. Temporal and spatial scales in phytoplankton ecology: mechanisms, methods, models, and management. Can. J. Fish. Aquat. Sci. 37:877–900.

Harris, G. P. 1986. Phytoplankton Ecology. London: Chapman and Hall.

Henrikson, L., H. G. Nyman, H. G. Oscarson, and J. A. E. Stenson. 1980. Trophic changes, without changes in the external nutrient loading. Hydrobiologia 68:257–263.

Hodgson, J. R. and J. F. Kitchell. 1987. Opportunistic foraging in largemouth bass (Micropterus salmoides). Amer. Midl. Nat. 118:323–336.

Hrbacek, J., M. Dvorakova, V. Korinek, and L. Prochazkova. 1961. Demonstration of the effect of the fish stock on the species composition of zooplankton and the intensity of metabolism of the whole plankton assemblage. Verh. Int. Ver. Theoret. Angew. Limnol. 14:192–195.

Kitchell, J. F. and S. R. Carpenter. 1987. Piscivores, planktivores, fossils, and phorbins. in: Predation: Direct and indirect impacts on aquatic communities, ed. W. C. Kerfoot and A. Sih. Hanover: University Press of New England.

Kratz, T. K., T. M. Frost, and J. J. Magnuson. 1987. Inferences from spatial and temporal variability in ecosystems: Analyses of long-term zooplankton data from a set of lakes. Am. Nat. 129:830–846.

Likens, G. E. 1985. An experimental approach for the study of ecosystems. J. Ecol. 73:381–396.

McQueen, D. J., J. R. Post, and E. L. Mills. 1986. Trophic relationships in freshwater pelagic ecosystems. Can. J. Fish. Aquat. Sci. 43:1571–1581.

Mills, E., J. Forney, and K. Wagner. 1987. Fish predation and its cascading effect on the Oneida Lake food chain. in: Predation: Direct and indirect impacts on aquatic communities, ed. W. C. Kerfoot and A. Sih. Hanover: University Press of New England.

Northcote, T. G. 1988. Fish in the structure and function of freshwater ecosystems: a top-down view. Can. J. Fish. Aquat. Sci. 45: in press.

Oksanen, L., S. D. Fretwell, J. Arruda, and P. Niemels. 1981. Exploitation ecosystems in gradients of primary productivity. Am. Nat. 118:240–261.

O'Neill, R. V., D. L. DeAngelis, J. B. Waide, and T. F. H. Allen. 1986. A Hierarchical Concept of Ecosystems. Princeton: Princeton University Press.

Pace, M. L. 1984. Zooplankton community structure, but not biomass, influences the phosphorus-chlorophyll a relationship. Can. J. Fish. Aquat. Sci. 41:1089–1096.

Paine, R. T. 1980. Food webs, linkage interaction strength, and community infrastructure. J. Anim. Ecol. 49:667–685.

Paine, R. T. 1981. Truth in ecology. Bull. Ecol. Soc. Amer. 62:256–258.

Peters, R. H. 1986. The role of prediction in limnology. Limnol. Oceanogr. 31:1143–1159.

Pielou, E. C. 1977. Mathematical ecology. New York: John Wiley and Sons.

Pitcher, T. J. and P. J. B. Hart. 1982. Fisheries ecology. Beckenham, Kent, England: Groom Helm Ltd.

Post, J. R. and D. J. McQueen. 1987. The impact of planktivorous fish on the structure of a plankton community. Freshwat. Biol. 17: in press.

Regier, H. A. and H. F. Henderson. 1973. Towards a broad ecological model of fish communities and fisheries. Trans. Am. Fish Soc. 102:56–72.

Riemann, B. and M. Sondergaard. 1986. Carbon dynamics of eutrophic, temperate lakes. Amsterdam: Elsevier Scientific Publishers.

Scavia, D., G. L. Fahnenstiehl, M. S. Evans, D. Jude, and J. T. Lehman. 1986. Influence of salmonid predation and weather on long-term water quality trends in Lake Michigan. Can. J. Fish. Aquat. Sci. 43:435–443.

Schindler, D. W. 1978. Factors regulating phytoplankton production and standing crop in the world's lakes. Limnol. Oceanogr. 23:478–486.

Shapiro, J. 1980. The importance of trophic-level interactions to the abundance and species composition of algae in lakes. in: Hypertrophic ecosystems, ed. J. Barica and L. Mur, 105–115. The Hague: Dr. W. Junk Publishing Co.

Shapiro, J., B. Forsberg, V. Lamarra, G. Lindmark, M. Lynch, E. Smeltzer, and G. Zoto. 1982. Experiments and experiences with biomanipulation. U.S. Environmental Protection Agency Report 600/3–82-096.

Shapiro, J. and D. I. Wright. 1984. Lake restoration by biomanipulation. Freshwat. Biol. 14:371–383.

Shugart, H. H., ed. 1978. Time series and ecological processes. Soc. Indust. and Appl. Math., Philadelphia.

Sissenwine, M. P. 1984. Why do fish populations vary? in: Exploitation of marine communities, ed. R. M. May, 59–94. New York: Springer-Verlag.

Sommer, U., Z. M. Gliwicz, W. Lampert, and A. Duncan. 1986. The PEG model of seasonal succession of planktonic events in fresh waters. Arch. Hydrobiol. 106:433–471.

Soutar, A. and J. D. Isaacs. 1974. History of fish populations inferred from fish scales in anaerobic sediments off California. Fish. Bull. 72:257–273.

Steele, J. H., ed. 1978. Spatial pattern in plankton communities. New York: Plenum.

Steele, J. H. 1984. Kinds of variability and uncertainty affecting fisheries. *in*: Exploitation of marine communities, ed. R. M. May, 245–262. New York: Springer-Verlag.

Strong, D. R. 1986. Density Vagueness: Abiding the variance in the demography of real populations. *in*: Community ecology, ed. J. Diamond and T. J. Case, 257–268. New York: Harper and Row.

Strub, P. T., T. Powell, and C. R. Goldman. 1985. Climatic forcing: Effects of El Nino on a small, temperate lake. Science 227:55–57.

Walters, C. 1986. Adaptive management of renewable resources. New York: MacMillan.

Warner, R. R. and P. L. Chesson. 1985. Coexistence mediated by recruitment fluctuations: A field guide to the storage effect. Am. Nat. 125:769–787.

West, D. L., H. H. Shugart, and D. B. Botkin (eds.). 1981. Forest Succession. New York: Springer-Verlag.

Wetzel, R. G. 1983. Limnology. Philadelphia: Saunders.

Part 4
Reports from Group Discussions

CHAPTER 9

Goals of the Discussion Groups

Stephen R. Carpenter, James F. Kitchell, William E. Neill, Donald Scavia, and Earl E. Werner

The first discussions at the workshop were organized around five topics encompassing major current issues in lake community ecology. Chairs and rapporteurs had at least two months to prepare for these discussions, in contrast to the later synthesis discussions (Chapters 15 and 16) which were more spontaneous. Each topic committee included knowledgeable scientists representing diverse backgrounds and perspectives. The general goal of the discussions was to determine recent advances, current problems, and research needs in each topic area. Chapters 10-14 were distilled from those far-reaching discussions and collectively define the horizon of research on complex interactions at the time of the workshop.

This chapter describes briefly the charges to the discussion groups, and the sorts of questions posed at the outset. The five discussion groups were:

1. *Food Web Effects*, including transmission of change through food webs, identification of important interactions (including criteria for importance), indirect effects as a special subset of complex interactions, and the role of time scale and temporal variability in food web processes.
2. *Size-Structured Interactions*, including regularities in community size structure and size dependent interactions, competitive bottlenecks, and the role of the ontogenic niche in food web structure.
3. *Habitat Interactions*, including the implications of habitat heterogeneity for complex interactions and specifically considering riparian/littoral, littoral/pelagial, and benthic/water column interactions.
4. *Microbial Interactions*, including the relationships of the microbial loop to traditional food web concepts, interactions of prokaryote and eukaryote communities, and the potential effects of engineered microbes on food web processes.
5. *Scale in the Design and Interpretation of Community Studies*, including implications of hierarchy theory, detecting change and imputing cause in studies of complex systems, experimental design where genuine replication is impossible, and implications of scaling assumptions for analyzing and interpreting complex systems.

Prior to the discussions, participants were given a list of questions for discussion. Some of these questions became foci for Chapters 10-14; others were not addressed by any group report. These questions are listed here both to record the starting points of the discussions and to provide a stimulus for further work.

1. *Aggregation of species into functional groups*: To what extent does the practice of aggregating species into functional groups or trophic levels limit our ability to resolve important interactions and construct predictive theory? Conversely, to what extent does exclusive focus at the species level limit our perspective on aquatic communities or increase research costs prohibitively? What information is lost by aggregating? What information is lost by working at the species level exclusively?

2. *Common or widespread linkage patterns*: Are there classes of interactions (linkage patterns among species or functional groups) that recur in many different types of communities? Are there linkage patterns among components that are common to systems at many levels of aggregation? What is the relative importance of linkage pattern vs interaction strength or flux rate in determining community or system behavior? If certain linkage patterns are widespread, should we expect similarities in system behavior despite differences in interaction strength or flux rate?

3. *Spatial and temporal heterogeneity*: How are complex interactions affected by spatial heterogeneity and by the scales of this heterogeneity? In what ways do complex interactions depend upon joint heterogenities in time and space? What are the roles of predictable, cyclic events vs irregular, aperiodic events in community dynamics? Are there short-lived events that have long-term consequences for complex interactions and community structure? What is the importance of the timing, synchrony, and/or asynchrony of events in determining complex interactions?

4. *Microbial and detrital processes in communities*: To what extent are microbial/detrital processes a consequence of the structure of the eukaryote community? Conversely, to what extent are microbial/detrital processes a determinant of eukaryote community structure? Does prokaryote biomass enter the traditional eukaryote food web to a significant extent? Does community structure determine if microbes are a source or a sink for carbon? To what extent do microbes and/or detrital processing govern resource supply ratios that determine the outcome of competition at upper trophic levels?

5. *Approaches to perturbation experiments*: What are the advantages and disadvantages of (1) experiments in which a key species or functional group is removed, and (2) response curve experiments employing a range of densities of key components? What are the relative merits of pulse and press experimental designs? How do the temporal and spatial scales of complex interactions influence the choice of experimental designs?

CHAPTER 10

Food Web Interactions in Lakes

Larry B. Crowder, Rapporteur, Ray W. Drenner,
Chair, W. Charles Kerfoot, Donald J. McQueen,
Edward L. Mills, Ulrich Sommer, Craig N. Spencer,
and Michael J. Vanni

Introduction

Why study food webs in lakes? From a basic research perspective, ecological studies of lake food webs provide distinct advantages over studies in many terrestrial systems (Lampert 1987). Lake food webs are composed of organisms with relatively fast population turnover rates which interact in a relatively closed system. These features allow us to readily observe the often rapid dynamics of these systems or to experimentally manipulate these food webs and quickly assess the system response. Enclosures, ponds, and whole-lake manipulations are extremely useful experimental tools that have allowed aquatic ecologists to test hypotheses on food web structure and function that would have been difficult or impossible to address in many terrestrial systems.

A second reason for studying lake food webs is to enhance water resource management. Lakes are extremely valuable as water sources for human use as well as natural habitats for aquatic organisms. Toxicants and other wastes related to human activities can have profound impacts on the structure and function of lake food webs, and much of the management oriented research has concentrated on the fate and effects of these pollutants. We have recently become aware that food web structure and function *itself* can influence the effects of nutrient additions and toxicants.

Understanding complex interactions within lake communities and their effects on energy flow and community structure is essential for effective management of lake systems. Most lakes are currently being managed to meet one goal or another: reductions in nutrient loadings to improve water quality; stocking or removing fish to meet fishery management objectives; managing to reduce toxic chemical loading or effects. However, most of these management actions are taken with insufficient understanding of community-level interactions.

Whether we seek to study complex interactions in lake food webs out of a desire to understand basic ecological processes and community structure or to acquire information for effective management, the questions demand sound science and a basic understanding of food web dynamics. An increase in understanding of direct and indirect effects in aquatic food webs is necessary to clarify the wide variety of system responses we may observe when lake communities are manipulated or perturbed.

Historical Perspectives on Lake Food Webs

Lake ecosystems have historically been studied primarily by people educated in one of two scientific disciplines, limnology or fishery biology, which have approached lakes from opposite extremes of the food web. Limnologists have concentrated on physical and chemical factors and their relationship to lower levels (e.g., phytoplankton and zooplankton), while fisheries biologists have studied fish populations with limited attention to the fishes' impact on lake food webs.

The earliest papers which began to suggest linkages between traditional limnology and fisheries were those which demonstrated effects of size selective planktivores on zooplankton community composition and size structure (Hrbacek et al. 1961; Hrbacek 1962; Brooks and Dodson 1965). These early insights on the role of planktivores in structuring the plankton community in the limnetic zone of lakes provided the impetus behind the development of much of our current understanding of the effects of fish on lake food webs.

The Complexity of Food Webs

Improving our understanding of lake food webs will not be a trivial task. Lake food webs are composed of both limnetic and littoral zone communities, each of which include hundreds of species ranging from microbes to plankton to insects to fish. These organisms have very different life histories; generation times of these populations vary from hours to years. The temporal dynamics of lake systems also depend upon seasonal and diel environmental cycles, and each group of organisms responds to these environmental changes differently based on its population dynamics. The response of different groups of organisms to environmental cues will thus exhibit different time dynamics. If these populations interact, then the temporal dynamics of one group may influence the dynamics of another, generating a complex pattern of temporal responses.

These complex temporal responses are nested within an equally complex spatial structure. Processes in the limnetic zone of lakes have profound interactions with littoral and benthic processes. Species undergo seasonal and diel migrations among these habitats, often contributing to the movement of materials or energy. Unfortunately, our current understanding of the coupling between littoral and limnetic regions and between benthic and limnetic regions is rudimentary (Lodge et al. this volume). In this chapter, we emphasize complex interactions in the limnetic zone.

This complexity is multiplied by the fact that resource use varies in important ways among taxa (e.g., bluegreens, diatoms, chrysophytes) and many species experience changes in their resource use patterns through ontogeny. Nutrient uptake kinetics are a function of phytoplankter size (Stein et al. this volume). Zooplankton, such as copepods, may be herbivorous in early instars, but omnivorous or carnivorous as adults (Chow-Fraser and Wong 1986; Williamson 1987). And while most all fish begin feeding upon plankton, many pass through a series of ontogenetic habitat or diet shifts as they grow (Werner and Gilliam 1984).

Another property of food webs that causes complexity is the nature of direct and indirect effects. In food web studies, predator-prey links are examples of direct interactions, whereas pure exploitative competition is an example of an indirect interaction, mediated through a shared resource pool. The classical keystone predator effect characterizes an interaction in which a predator has an indirect beneficial effect on a suite of inferior competitors by depressing the abundance of a superior competitor (Paine 1966; Lubchenco 1978; Paine 1980). A related effect develops in cascading trophic interactions when a top consumer benefits prey two trophic links lower in the food web because it reduces the abundance of an intermediate consumer (Hurlbert et al. 1971; Abrams 1984; Carpenter et al. 1985).

Important indirect effects can arise when two predators depend on two competitors as prey. For example, if fish suppress large zooplankton, small zooplankton have more food available and thus increase in density. Increased abundance of small zooplankton could have beneficial effects for an invertebrate predator that consumes small zooplankton. Interactions of this kind between fish and invertebrate predators were termed complementary feeding niches by Dodson (1970). Dodson's 1970 paper inspired a major theoretical contribution by Levine (1976), who pointed out how competing consumers may show mutualistic effects. Mutualistic effects among invertebrate predators have been demonstrated for *Chaoborus* (Gigiere 1979, cited in Paine 1980; Neill this volume) and *Mesocyclops* (Kerfoot and DeMott 1984; Kerfoot 1987). At intermediate densities, fish can reduce the dominant competitor (*Daphnia*), allowing more food for small-bodied zooplanton and also allowing invertebrate predators to increase in number. At high fish densities, invertebrate predators are suppressed by predation, while at low fish densities *Daphnia* reduce the abundance of small zooplankton and thus can reduce food levels for invertebrate predators. These interactions create optimum conditions for invertebrate predators at intermediate fish densities (Kerfoot 1987).

Indirect effects in food webs involving invertebrate predators lead to a rich fabric of potential interactions which have not yet been carefully explored. Neill (1985) demonstrated these effects experimentally by reducing

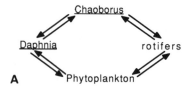

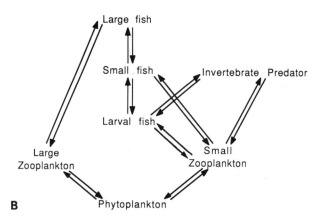

A

B

FIGURE 10.1. (A) Food web with indirect effects of an invertebrate predator, *Chaoborus*, on *Daphnia* (based on Neill 1985). Arrows pointing up relate to material or energy flow. Arrows pointing down reflect predatory control of prey. (B) Food web with indirect effects involving an invertebrate predator and different onto-genetic stages of fish.

Daphnia populations which allowed more food for rotifers and thereby increased survival of early *Chaoborus* instars which feed on rotifers. Later instars of these invertebrate predators then prevented *Daphnia* from re-covering (Fig. 10.1A). The direct and indirect effects in food webs involving invertebrate predators and fish become particularly interesting when one considers size-structured interactions with fish as well (Fig. 10.1B). This set of interactions between invertebrate predators and fish can have in-teraction signs of (-, +), (-, -) and (+, -) as fish move through their on-togeny (cf. Stein et al. this volume). These complex interactions among invertebrate predators and fish have been documented from the littoral zone as well (Crowder and Cooper 1982).

Behavioral indirect effects are particularly difficult to address in the context of food webs, because behavioral responses often involve move-ment to different habitats, and food webs do not incorporate such spatial information. Behavioral effects are important and apparently widespread

(Sih 1980; Sih et al. 1985; Werner et al. 1983; Mittelbach 1984; Mittelbach and Chesson 1987). Responses of prey, such as movement to refuges or changing activity levels in the presence of predators, may cause changes in food web linkages.

Current Status of Food Web Studies in Lakes

Food Web Analysis

Food webs as pictured in standard textbooks in ecology, limnology or fisheries biology often do not give a realistic picture of the structure and complexity of food webs. Even webs that appear to be representations of real ecological systems, including some of those subjected to elaborate theoretical analyses (cf. Cohen 1978; Pimm 1980, 1982), are really only caricatures of actual webs. Like all caricatures, these webs exaggerate those characters that the artist chooses to emphasize while some characters may be completely omitted. Limnologists usually detail nutrient pathways and components at the bottom end of aquatic food webs, while aggregating upper trophic-level components into one or two compartments (Fig. 10.2A). By contrast, fisheries biologists often exhibit a different viewpoint; fish are detailed to species but phytoplankton species are rarely distinguished and nutrients are not considered at all (Fig. 10.2B). Both groups have yet to consider fully the importance of the microbial component of freshwater food webs (Porter et al. this volume), in which omnivory is the rule and organisms can sometimes be photosynthetic and sometimes heterotrophic.

Further, most food webs are presented as static. Most are drawn from descriptions of the connections among food web components at one time, in one location, or they are some kind of seasonal average web which may never exist in a particular time or place. In fact, one might expect food web structure and function to be highly dynamic, at least in seasonal environments (Mills and Forney this volume). We have been trying to achieve an understanding of these highly dynamic food web processes by looking at the equivalent of blurred 8 by 10 glossies. We clearly need to move on to "super-8" or "video" representations of these dynamic systems.

Seasonal succession in food webs is probably driven by physical/chemical environmental changes interacting with the seasonal ontogeny of organisms and species interactions. Environmental variability contributes both to setting initial conditions for seasonal phytoplankton succession (Sommer 1985a,b; Sommer et al. 1986) and to large-scale recruitment variability in fishes (Mills et al. 1987; Mills and Forney this volume). In order to move beyond the "snapshot" phase in our understanding of food web dynamics, we will need improved understanding of these seasonal effects.

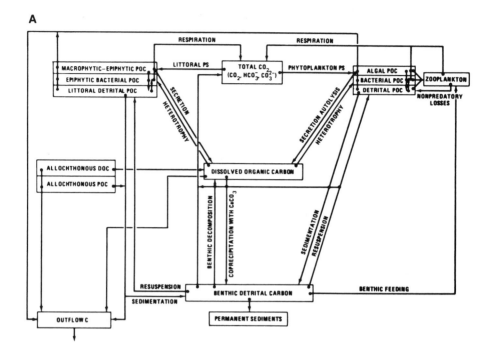

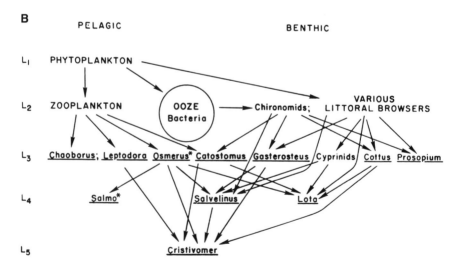

FIGURE 10.2. (A) Major flow pathways of organic carbon in Lawrence Lake, Michigan, from a limnological perspective (from Wetzel and Rich 1973). (B) Food web for Moosehead Lake, Maine, from a fishery biology perspective (from Frey 1963. © The Board of Regents of the University of Wisconsin).

Year to year variation in food web structure and function is sensitive to initial conditions, such as yearly variation in temperature, water levels, thermal stratification, nutrient inputs, and their influence on biotic processes. Yet, successional sequences in north temperate regions fall into predictable scenarios (Sommer et al. 1986; Lampert 1987). As Carpenter (this volume) and Mills and Forney (this volume) noted, year class formation in fish and its variability are important because of the long life span of fishes and the resultant long influence on food web structure and function.

Strong Interactions and Aggregation in Food Webs

These complexities would be cause for despair except, as noted by Paine (1980), understanding real food webs can be simplified even though they have a large number of components with an even larger number of connections. Paine's point is that communities are often dominated by just a few strong interactions which can be experimentally examined (Fig. 10.3). Further, species may be aggregated into functional groups based on ecological similarities. Aggregation is a tricky subject, best based on objective protocols rather than arbitrary decisions (Frost et al. this volume). In large part, it seems that the appropriate or acceptable degree of aggregation of food web components will depend upon the question being asked.

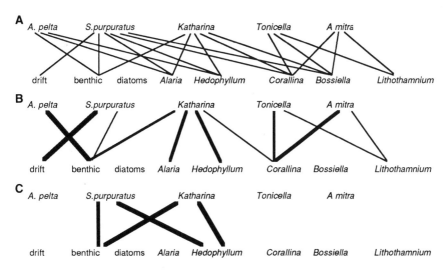

FIGURE 10.3. Three conceptually and historically different approaches to depicting trophic relationships in rocky intertidal communities, illustrated for the same set of species. The connectedness web (A) is based on observation, the energy flow web (B) on measurements and literature values, and the functional one (C) on controlled manipulation (from Paine 1980).

Within the context of a particular question, how does one aggregate specific food web components into functional groups? It seems that the ultimate criterion is that components included in the same functional group should be substitutable. By this we mean that some species shifts can occur in food webs without major changes in food web structure or function. In order to meet this criterion, one must consider constraints on substitutability in the food web both from below and from above. For example, aggregating planktivorous fish into functional groups based on their feeding modes and the zooplankton resources they consume might lead to the inclusion of bluegills and fathead minnows in the same functional group (visually feeding planktivores). However, if these species were aggregated based on vulnerability to visually feeding size selective piscivores, such as largemouth bass, they probably would not be placed in the same functional group because bluegills are spiny deep-bodied fish that are less vulnerable than fathead minnows. Obviously, aggregation will be a risky trade-off between added simplicity and the potential for misleading results and should be considered carefully. Paine (1980) noted that aggregation into trophic levels may well be unreasonable in aquatic systems because so many organisms are omnivorous and because many shift trophic roles through ontogeny (cf. Werner and Gilliam 1984).

Top-Down vs Bottom-up Effects

Food webs have been viewed from a variety of perspectives. If the emphasis of the researchers was on energy flow, the arrows often point up. If the emphasis was on predator effects, the control arrows may point down. In fact, at each trophic level, resources from the trophic level below will enhance production, while predator removal from above will act to alter community structure, productivity, or biomass.

Top-down effects of piscivores on planktivores apparently can reduce planktivory, leading to increased zooplankton size, reduced chlorophyll, and clearer water (Edmondson and Litt 1982; Mills et al. 1987), even in the largest US lake (speculation by Brooks 1969; documented by Scavia et al. 1986; Kitchell and Crowder 1986). Recently, a number of experimental tests of this idea have been performed (Shapiro et al. 1975; Henrikson et al. 1980; Shapiro and Wright 1984; Spencer and King 1984; Kerfoot and DeMott 1984; Levitan et al. 1985; Hambright et al. 1986; Carpenter et al. 1987).

Top-down effects often include qualitative community shifts where populations of selected prey are reduced (direct effects of predation) and less utilized prey increase (indirect effects). This may complicate or mask the clear cascading effects previously described. For example, large (e.g. adult) visually feeding zooplanktivores selectively remove large bodied zooplankton and enhance populations of small zooplankton (Brooks and Dodson 1965; Hall et al. 1976; Zaret 1980; Kerfoot and Sih 1987). But smaller planktivores (e.g. age-0 fish) selectively remove small and intermediate sized zooplankton, and, if abundant, can increase the relative abundance of large zooplankton (Mills and Forney this volume; Crowder et al. 1987). Thus, fish effects on zooplankton community structure depend on predator size. In littoral habitats, visual feeders suppress large, active prey and enhance smaller, more cryptic forms (Crowder and Cooper 1982). Filter feeding omnivorous fish suppress slow zooplankters (e.g. those with low escape ability) and large algae and enhance the fast zooplankton and small algae (Drenner et al. 1986). Zooplankton grazers can suppress small phytoplankton and lead to shifts toward large forms (Lynch and Shapiro 1981; Bergquist et al. 1985). Finally, the impact of piscivore predation on planktivores may be reduced in some systems where much of the planktivore population is not available to the predators due to their large body size (Jenkins 1979; Noble 1981; Hambright et al. 1986).

Further, predation can have behavioral (nonlethal) as well as lethal effects on prey. Increasing numbers of examples are documenting that many small fishes alter their habitat choice and/or activity levels in the presence of significant predation risk (cf. Werner and Gilliam 1984; Power 1984, 1987; Mittelbach and Chesson 1987). Small centrarchids often occupy littoral vegetation as a refuge from predation by largemouth bass, sometimes leading to increased competition (Mittelbach 1984; Mittelbach and Chesson 1987).

Planktivory clearly can alter zooplankton community structure, but it is still unclear whether top-down effects will lead to sustained shifts in biomass of zooplankton or phytoplankton. In systems where ambient planktivore densities are sufficient to reduce large grazers, reductions in planktivorous fish densities in short term experiments can lead to increases in total zooplankton biomass and/or mean zooplankter biomass along with reductions in phytoplankton biomass (McQueen and Post 1984; Spencer and King 1984; Post and McQueen 1987; Vanni 1987a,b). In lakes where fish biomass is low, removal may have little effect, although fish additions can lead to the expected effects (DeMott and Kerfoot 1982; Levitan et al. 1985). Longer-term whole-lake manipulations (Henrikson et al. 1980; Carpenter et al. 1987) noted similar patterns in biomass dynamics as well as reductions in primary productivity when planktivory was severely reduced.

Many of the experiments cited involved fairly extreme manipulations of predation (e.g., fish vs no fish; high densities of piscivores relative to planktivore biomass). It is unclear whether these patterns can be sustained; even the long-term, whole-lake manipulations (Henrikson 1980; Carpenter et al. 1987) have not included a time period long enough to allow a numerical response of the various piscivore and planktivore populations to manipulations at the top of the food web.

Support for the notion that nutrient effects ripple up the web, influencing biomass at each level, is strong (McQueen et al. 1986). A common fish management tool, the morphoedaphic index (Ryder 1965), makes reasonable predictions of fish yields based on an index of lake morphometry and nutrient conditions. Documentation that bottom-up effects could influence food web structure is limited, but a couple of interesting observations suggest that these effects are possible. The first is that variation in intensity of spring mixing in lakes can yield different initial nutrient conditions. Thus, seasonal algal succession can begin with different species compositions (Sommer 1985). Second, Smith (1983) has shown that shifts in nutrient ratios (N/P) can lead to competitive shifts in phytoplankton community composition which may be differentially useful to zooplankton grazers.

The relative effects of top-down vs bottom-up forces in food web structure and function are unclear—both seem to have the potential for strong effects. But much additional research will be necessary to estimate the balancing rates that determine the relative importance of these two vectors. Even at one of the most studied nodes in the food web, *Daphnia*, we have strong competing hypotheses. Clear water periods which are regularly observed in many meso- and eutrophic lakes can be caused by *Daphnia* grazing (Lampert et al. 1986). Regression analysis of chlorophyll *a* and zooplankton biomass has shown that cladoceran body size is a significant factor explaining residual variation in the total phosphorus-chlorophyll relationship (Pace 1984). When *D. pulex* were introduced into enclosures in a eutrophic lake, phytoplankton abundance decreased by an order of magnitude (Vanni 1986). In contrast, the loss of large cladocerans such as *Daphnia* often coincides with an increase in blue-green algal filaments (Spencer and King 1984, 1986) which can lead to increased rates of food rejection by *Daphnia* and thus higher respiration rates, inhibiting larger *Daphnia* (Webster and Peters 1978, Porter and McDonough 1984).

Research Needs in Lake Food Webs

Lake Trophic Status and Top-Down Effects

One issue that still remains controversial is the effect of lake trophic status on top-down effects. McQueen et al. (1986) have hypothesized that top-down effects in eutrophic lakes are strong from piscivore to planktivore, somewhat weaker for planktivore to zooplankton, and have little impact for zooplankton on phytoplankton. Their model predicts that top-down effects are not as well buffered in oligotrophic lakes causing zooplankton-phytoplankton interactions to be stronger in oligotrophic systems.

However, experimental manipulations of organisms at several trophic levels suggest that top-down effects can also be very strong in eutrophic systems. When planktivorous fish were added to eutrophic systems containing no fish, *Daphnia* declined dramatically and phytoplankton increased (Hrbacek 1962; Losos and Hetesa 1973; Lynch 1979; Lynch and

Shapiro 1981; Spencer and King 1984). Schoenberg and Carlson (1984) directly manipulated zooplankton abundance and community structure in enclosures in a hypereutrophic lake, and found that enclosures with abundant *Daphnia galeata mendotae* had much lower phytoplankton abundance than those with *Bosmina*.

Two of these experiments simultaneously manipulated nutrients and planktivores and compared their effects on phytoplankton community structure and biomass. Adding fish to fishless enclosures in eutrophic systems had much larger effects on the phytoplankton than did adding nutrients to these systems (Losos and Hetesa 1973; Lynch and Shapiro 1981). Two additional experiments show that top-down effects may be enhanced in eutrophic systems relative to oligotrophic systems. *Chaoborus* had greater effects on herbivorous zooplankton when nutrients were added to enclosures in an oligotrophic lake (Neill and Peacock 1980). Similarly, *Daphnia* reduced phytoplankton biomass an order of magnitude more in enclosures placed in a eutrophic lake than in enclosures placed in an oligo-mesotrophic lake (Vanni 1986).

Methodological Questions

It is still unclear to what extent the observed top-down effects in enclosure and mesocosm experiments might be influenced by initial conditions. Some experiments begin by introducing fish to fishless systems; generally the treatment effects in these experiments have been rapid and strong. If predators are, instead, removed from the lake or enclosure, the experiment must be long enough to allow colonization and numerical responses of prey previously suppressed by the predator.

Most of the current studies of top-down effects involve extreme fish manipulations (all or none) and often at unrealistic densities (the so-called sledgehammer manipulation). While we agree that it is reasonable to explore the extremes of predator density, particularly in early experiments, one must also be sure that the treatments used are ecologically realistic. The use of extreme manipulations may often achieve statistically significant results of little ecological significance. Using extreme fish biomass treatments in mesocosms to examine the effects of planktivores on their prey can lead to weight loss and even death in stocked fishes (Threlkeld 1987). Stressed fish that are losing weight during experiments are likely to release phosphorus at disproportionate rates (Kitchell et al. 1975), so that extreme top-down manipulations may also be nutrient addition experiments (cf. Threlkeld 1987).

This suggests that interpretations of so-called sledgehammer experiments could be problematic. If the treatments are so extreme that they totally reconfigure the system, the results may not be relevant to the original lake system. Another concern about these two-level experiments (high-low or some-none) is that if the system response is actually nonlinear (Fig.

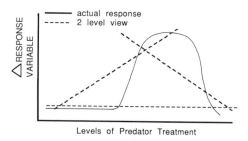

FIGURE 10.4. Misleading results possible from performing a two level experiment on a system with nonlinear dynamics. Depending on the two levels chosen for the experiment, one could find positive, negative, or no relationship.

10.4), one can get positive correlations, negative correlations, or no correlations out of what is actually a nonlinear response, depending on the treatment levels chosen. Gradient experiments may help solve this problem. Neill's (1985) gradient experiment on *Daphnia* removal clarifies important nonlinearities involving rotifers and *Chaoborus* that simply would not have been usefully explored in a two-level experiment. Removal of a large percentage of the *Daphnia* populations in enclosures in July allowed rotifers to increase (Fig. 10.5). This increase in rotifers allowed increased survival of early instars of *Chaoborus*, which became abundant and subsequently supressed the recovery of *Daphnia* in the enclosures.

Spatial and Temporal Considerations in Food Webs

Our understanding of lake food webs primarily derives from observations and experiments in the epilimnion of north temperate clear-water lakes. Our understanding of ties to the littoral or benthic webs is, at best, rudimentary. We expect that the littoral and benthic habitats have important effects on pelagic webs both in terms of nutrient dynamics and food web linkages. Through their ontogeny, fish occupy and use all of these habitats. In this way, fish may be considered to be important spatial vectors. As they develop and move through various habitats, their top-down effects shift from one food web to another. They may also contribute to nutrient transfer among these systems. Lodge et al. (this volume) have discussed this coupling problem at length and agree that it merits further research.

Temporal variation in food web dynamics also requires increased attention. Given the fact that relatively long-lived organisms seem to have a strong influence on system structure and function (Carpenter, this volume, Mills and Forney, this volume), it is clear that we need additional research on the mechanisms underlying fish recruitment variation in lake communities. Both abiotic and biotic factors significantly affect fish recruitment, and the recruitment dynamics of fish probably depend on their own web of complex interactions. These short term dynamics lead to var-

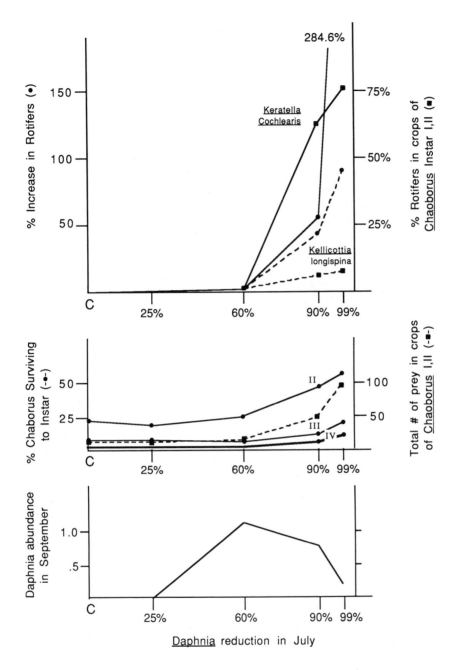

FIGURE 10.5. Direct and indirect interactions in a *Daphnia* removal experiment (based on Neill 1985). In the top panel, circles represent % increase in abundance of two species of rotifers and squares represent % rotifers in *Chaoborus* diets. In the middle panel, circles represent % *Chaoborus* survival and squares represent # of prey in *Chaoborus* diets. The bottom panel shows relative abundance of *Daphnia* in September for treatments with different *Daphnia* removal rates 2 months earlier.

iable recruitment which results in long term effects in lake food webs and will require detailed study.

On a longer time scale, we know little about the effects of coevolution on food web structure and function. Lakes and their food webs vary in age from millions to thousands to tens of years. In old lakes, natural selection seems to have resulted in species assemblages and food webs that are highly diverse and specialized. But the recent history of the Nile perch invasion in Lake Victoria (Barel et al. 1985) suggests that these systems can be highly vulnerable to the addition of an exotic piscivore. In the U.S., lake ecosystems range from 10,000 yr old (many with introduced species) to reservoirs only a few years old with artificially assembled faunas. Lakes with a long coevolutionary history may have more complex webs with greater niche specificity and more redundancy. In young systems, particularly in reservoirs, we often see assemblages of generalists with little opportunity for coevolution, though an evolutionary perspective will still be useful toward understanding these systems (Werner 1980).

Approaches to the Study of Lake Food Webs

The study of complex interactions in lake food webs will require a multiplicity of approaches. Given the strong propensity among ecologists to promote "bandwagons" or particular approaches to particular problems, we feel that we must emphasize the power of a more diverse approach. One only need examine the contents of the plenary papers in this book to note that what we now understand about lake food webs derives from long-term empirical studies, experiments (laboratory, field mesocosm, whole lake), and modeling efforts. No one approach will lead to a complete understanding of how food webs function. Clearly, the exposition of alternate hypotheses and examining interactions from a variety of perspectives are most likely to lead toward rapid progress (Kitchell et al. this volume).

As Persson et al. (this volume) note, traditional education in limnology focuses on physical and chemical factors that constrain energy and nutrient dynamics in lakes; little emphasis is given to benthic animals or fish. By contrast, traditional training in fisheries biology focuses on ichthyology, population dynamics and management, with little emphasis being given to other components of the aquatic food web except those utilized by fish. In order to address complex interactions in lake food webs, education of aquatic ecologists will have to be broadened to include much of the information formerly included in the separate disciplines of limnology and fisheries. Rigler (1982) detailed the history of the schism between limnology and fisheries and noted that demands for integration in understanding of aquatic systems for both basic research and applied interests would require some sort of synthesis of these two fields. Given that current lake management institutions and fishery management institutions are often separate and that communication between these groups has traditionally been limited, effective management of lake ecosystems may require some institutional realignments.

The evolutionary perspective is probably more common among people educated in traditional limnology and aquatic ecology than it is among fishery biologists (cf. Kerr 1980; Werner 1980). People interested in managing particular lake processes or components of lake food webs are often more concerned with short term (ecological time scale) dynamics than with evolutionary time scales. Nevertheless, these lake processes and food web components result from evolutionary processes acting on these systems. Even for communities in relatively young systems (e.g. reservoirs), understanding the evolutionary history of the organisms and the ecological context under which they evolved should provide useful information, even to managers interested in short term dynamics. To date, ecologists have not generally gotten this idea across to managers of these systems (Kerr 1980; Werner 1980). Recent examples of fishery declines following the introduction of *Mysis* illustrate the potential consequences of management activities carried out without adequate understanding of the ecology and evolutionary history of introduced species (Lasenby et al. 1986).

Complex interactions in lake food webs provide a difficult and challenging frontier for community ecologists. Further, the potential management utility of some of these dynamics (e.g., biomanipulation) is large. We urge caution in moving too rapidly toward management with food web manipulations. As illustrated in this chapter, our current understanding of top-down and bottom-up effects on lake food webs is in an early stage of development. We continue to encounter surprises in our experimental manipulations and given the rich fabric of potential indirect effects, we doubt that we will soon be able to write reliable prescriptions for biomanipulation. This is not to discourage field trials of this approach, but to suggest that much additional research will be necessary before this approach becomes a standard tool in the lake manager's kit. Similarly, traditional control of nutrient inputs as a water quality management strategy, without consideration of past or present alterations in the food web may also lead to surprises. Future research and management must depend on an ecosystem perspective so that important interactions will not be overlooked.

References

Abrams, P. A. 1984. Foraging time optimization and interactions in food webs. Am. Nat. 124:80–96.

Barel, C. D. N., R. Dorit, P. H. Greenwood, G. Fryer, N. Hughes, P. B. N. Jackson, H. Kawanabe, R. H. Lowe-McConnell, M. Nagoshi, A. J. Ribbink, E. Trewavas, F. Witte and K. Yamaoka. 1985. Destruction of fisheries in Africa's lakes. Nature 315:19–20.

Bergquist, A. M., S. R. Carpenter and J. C. Latino. 1985. Shifts in phytoplankton size structure and community composition during grazing by contrasting zooplankton assemblages. Limnol. Oceanogr. 30:1037–1045.

Brooks, J. L. 1969. Eutrophication and changes in the composition of zooplankton. in: Eutrophication: Causes, consequences, correctives. 236–255. Washington D.C.: National Academy of Sciences.

Brooks, J. L. and S. I. Dodson. 1965. Predation, body size, and composition of plankton. Science 150:28–35.

Carpenter, S. R., J. F. Kitchell and J. R. Hodgson. 1985. Cascading trophic interactions and lake productivity. Bioscience 35:634–639.

Carpenter, S. R., J. F. Kitchell, J. R. Hodgson, P. A. Cochran, J. J. Elser, M. M. Elser, D. M. Lodge, D. Kretchmer, X. He and C. N. von Ende. 1987. Regulation of lake primary productivity by food-web structure. Ecology 68:1863–1876.

Chow-Fraser, P. and C. K. Wong. 1986. Dietary change during development in the freshwater calanoid copepod *Epischura lacustris* Forbes. Can. J. Fish. Aquat. Sci. 43:938–944.

Cohen, J. 1978. Food webs and niche space. Princeton: Princeton University Press.

Crowder, L. B. and W. E. Cooper. 1982. Habitat structural complexity and the interaction between bluegills and their prey. Ecology 63:1802–1813.

Crowder, L. B., M. E. McDonald and J. A. Rice. 1987. Understanding recruitment of Lake Michigan fishes: The importance of size-based interactions between fish and zooplankton. Can. J. Fish. Aquat. Sci. 44 (Suppl. 2):141–147.

DeMott, W. R. and W. C. Kerfoot. 1982. Competition among cladocerans: nature of the interaction between *Bosmina* and *Daphnia*. Ecology 63:1949–1966.

Dodson, S. I. 1970. Complementary feeding niches sustained by size-selective predation. Limnol. Oceanogr. 15:131–137.

Drenner, R. W., S. T. Threlkeld and M. D. McCracken. 1986. Experimental analysis of direct and indirect effects of an omnivorous filter-feeding clupeid on plankton community structure. Can. J. Fish. Aquat. Sci. 43:1935–1945.

Edmondson, W. T. and A. Litt. 1982. *Daphnia* in Lake Washington. Limnol. Oceanogr. 27:272–293.

Frey, D. G. 1963. Limnology in North America. Madison: University of Wisconsin Press.

Hall, D. J., S. T. Threlkeld, C. Burns and P. H. Crowley. 1976. The size-efficiency hypothesis and the size structure of zooplankton communities. Ann. Rev. Ecol. Syst. 7:177–208.

Hambright, K. D., R. J. Trebatoski, R. W. Drenner and D. Kettle. 1986. Experimental study of the impacts of bluegill (*Lepomis macrochirus*) and largemouth bass (*Micropterus salmoides*) on pond community structure. Can. J. Fish. Aquat. Sci. 43:1171–1176.

Henrikson, L., H. G. Nyman, H. G. Oscarson and J. A. E. Stenson. 1980. Trophic changes without changes in external nutrient loading. Hydrobiologia 68:257–263.

Hrbacek, J., M. Dvorakova, V. Korinek and L. Prochazkova. 1961. Demonstration of the effect of the fish stock on the species composition of zooplankton and the intensity of metabolism of the whole plankton assemblage. Verh. Int. ver. Theoret. Angew. Limnol. 14:192–195.

Hrbacek, J. 1962. Species composition and the amount of zooplankton in relation to fish stock. Rozpravy Ceshoslovenske Akademie Ved Rada Matematickych a Prodnich Ved 72:1–116.

Hurlburt, S. H., J. Zedler and D. Fairbanks. 1971. Ecosystem alteration by mosquitofish (*Gambusia affinis*) predation. Science 178:639–641.

Jenkins, R. M. 1979. Predator-prey relations in reservoirs. *in*: Predator-prey systems in fisheries management, ed. R. H. Stroud and H. Clepper, 123–134. Washington D.C.: Sport Fishing Institute.

Kerfoot, W. C. 1987. Cascading effects and indirect pathways. *in*: Predation: Direct and indirect impacts on aquatic communities, ed. W. C. Kerfoot and A. Sih, 57–70. Hanover: University Press of New England.

Kerfoot, W. C. and W. R. DeMott. 1984. Food web dynamics: dependent chains and vaulting. *in*: Trophic interactions within aquatic ecosystems, ed. D. G. Meyers and J. R. Strickler, 347–382. Boulder: Westview Press.

Kerfoot, W. C. and A. Sih, eds. 1987. Predation: Direct and indirect impacts on aquatic communities. Hanover: University Press of New England.

Kerr, S. R. 1980. Niche theory and fisheries ecology. Trans. Amer. Fish. Soc. 109:254–257.

Kitchell, J. F., J. F. Koonce and P. S. Tennis. 1975. Phosphorus flux through fishes. Verh. Internat. Verein. Limnol. 19:2478–2484.

Kitchell, J. F. and L. B. Crowder. 1986. Predator-prey interactions in Lake Michigan: Model predictions and recent dynamics. Env. Biol. Fish. 16:205–211.

Lampert, W. 1987. Predictability in lake ecosystems: The role of biotic interactions. *in*: Ecological studies, Vol. 61, ed. E.-D. Schulze and H. Zwolfer, 333–346. Berlin: Springer-Verlag.

Lampert, W., W. Fleckner, H. Rai and B. E. Taylor. 1986. Phytoplankton control by grazing zooplankton: Study on the spring clear-water phase. Limnol. Oceanogr. 31:487–490.

Lasenby, D. C., T. G. Northcote and M. Furst. 1986. Theory, practice and effects of *Mysis relicta* introductions of North American and Scandinavian Lakes. Can. J. Fish. Aquat. Sci. 43:1277–1284.

Levine, S. H. 1976. Competitive interactions in ecosystems. Am. Nat. 110:903–910.

Levitan, C., W. C. Kerfoot and W. R. DeMott. 1985. Ability of *Daphnia* to buffer trout lakes against periodic nutrient inputs. Verh. Internat. Verein. Limnol. 22:1–7.

Losos, B. and J. Hetesa. 1973. The effect of mineral fertilization and of carp fry on the composition and dynamics of plankton. Hydrobiological Studies 3:173–217.

Lubchenco, J. 1978. Plant species diversity in a marine rocky intertidal community: Importance of herbivore food preference and algal competitive abilities. Am. Nat. 112:23–39.

Lynch, M. 1979. Predation, competition, and zooplankton community structure: An experimental study. Limnol. Oceanogr. 24:253–72.

Lynch, M. and J. Shapiro, 1981. Predation, enrichment, and phytoplankton community structure. Limnol. Oceanogr. 26: 86–102.

McQueen, D. J. and J. R. Post. 1984. Effects of planktivorous fish on zooplankton, phytoplankton and water chemistry. Lake and Reservoir Management,

Proceedings of the Fourth Annual Conference, NALMS, McAfee, New Jersey.

McQueen, D. J., J. R. Post and E. L. Mills. 1986. Trophic relationships in freshwater pelagic ecosystems. Can. J. Fish. Aquat. Sci. 43:1571–1581.

Mills, E. L., J. L. Forney and K. J. Wagner. 1987. Fish predation and its cascading effect on the Oneida Lake food chain. *in*: Predation: Direct and indirect impacts on aquatic communities, ed. W. C. Kerfoot and A. Sih, 118–131. Hanover: University Press of New England.

Mittelbach, G. G. 1984. Predation and resource partitioning in two sunfishes (Centrarchidae). Ecology 65:499–513.

Mittelbach, G. G. and P. L. Chesson. 1987. Predation risk: Indirect effects on fish populations. *in*: Predation: Direct and indirect impacts on aquatic communities, ed. W. C. Kerfoot and A. Sih, 315–332. Hanover: University Press of New England.

Neill, W. E. 1985. The effects of herbivore competition upon the dynamics of *Chaoborus* predation. Arch. Hydrobiol. 21: 483–491.

Neill, W. E. and A. Peacock. 1980. Breaking the bottleneck: Interactions of invertebrate predators and nutrients in oligotrophic lakes. *in*: Evolution and Ecology of Zooplankton Communities, ed. W. C. Kerfoot, 715–724. Hanover: University Press of New England.

Noble, R. L. 1981. Management of forage fishes in impoundments of the southern United States. Trans. Amer. Fish. Soc. 110:738–750.

Pace, M. L. 1984. Zooplankton community structure, but not biomass, influences the phosphorus-chlorophyll *a* relationship. Can. J. Fish. Aquat. Sci. 41:1089–1096.

Paine, R. T. 1966. Food web complexity and species diversity. Am. Nat. 100:65–75.

Paine, R. T. 1980. Food webs, linkage interaction strength, and community infrastructure. J. Anim. Ecol. 49:667–685.

Pimm, S. L. 1980. Properties of food webs. Ecology 61:219–225.

Pimm, S. L. 1982. Food webs. London: Chapman and Hall.

Porter, K. G. and R. McDonogough. 1984. The energetic cost of response to bluegreen algal filaments by cladocerans. Limnol. Oceanogr. 29:365–369.

Post, J. R. and D. J. McQueen. 1987. The impact of planktivorous fish on the structure of a plankton community. Freshwater Biology 17:79–89.

Power, M. E. 1984. Depth distribution of armored catfish: Predator induced resource avoidance? Ecology 65:523–528.

Power, M. E. 1987. Predator avoidance by grazing fishes in temperate and tropical streams: Importance of stream depth and prey size. *in*: Predation: Direct and indirect impacts on aquatic communities, ed. W. C. Kerfoot and A. Sih, 333–351. Hanover: University Press of New England.

Rigler, F. H. 1982. The relation between fisheries management and limnology. Trans. Amer. Fish. Soc. 111:121–132.

Ryder, R. A. 1965. A method for estimating the fish production of north-temperate lakes. Trans. Amer. Fish. Soc. 94: 214–218.

Scavia, D., G. L. Fahnenstiel, M. S. Evans, D. Jude and J. T. Lehman. 1986. Influence of salmonine predation and weather on long-term water quality trends in Lake Michigan. Can. J. Fish. Aquat. Sci. 43:435–443.

Schoenberg, S. A. and R. E. Carlson. 1984. Direct and indirect effects of zooplankton grazing on phytoplankton in a hypereutrophic lake. Oikos 42:291–302.

Shapiro, J., V. Lamarra and M. Lynch. 1975. Biomanipulation: An ecosystem approach to lake restoration. *in*: Proceeding of a symposium on water quality management through biological control, ed. P. L. Brezonik and J. L. Fox, 85–96. Gainesville: University of Florida.

Shapiro, J. and D. I. Wright. 1984. Lake restoration by biomanipulation: Round Lake, Minnesota, the first two years. Freshwater Biol. 14:371–383.

Sih, A. 1980. Optimal behavior: Can foragers balance two conflicting demands? Science 210:1041–43.

Sih, A., P.Crowley, M. McPeek, J. Petranka and K. Strohmeier. 1985. Predation, competition, and prey communities: A review of field experiments. Ann. Rev. Ecol. Syst. 16: 269–305.

Smith, V. H. 1983. Low nitrogen to phosphorus ratios favor dominance by bluegreen algae in lake phytoplankton. Science 221:669–671.

Sommer, U. 1985a. Comparison between steady state and non-steady state competition: Experiments with natural phytoplankton. Limnol. Oceanogr. 30:335–346.

Sommer, U. 1985b. Seasonal succession of phytoplankton in Lake Constance. BioScience 35:351–357.

Sommer, U., Z. M. Gliwicz, W. Lampert and A. Duncan. 1986. The PEG model of seasonal succession of planktonic events in freshwaters. Arch. Hydrobiol. 106:433–471.

Spencer, C. N. and D. L. King. 1984. Role of fish in regulation of plant and animal communities in eutrophic ponds. Can. J. Fish. Aquat. Sci. 41:1851–1855.

Spencer, C. N. and D. L. King. 1986. Regulation of blue-green algal buoyancy and bloom formation by light, inorganic nitrogen, CO_2, and trophic level interactions. Hydrobiologia 144:183–192.

Threlkeld, S. T. 1987. Experimental evaluation of trophic-cascade and nutrient-mediated effects of planktivorous fish on plankton community structure. *in*: Predation: Direct and indirect impacts on aquatic communities, ed. W. C. Kerfoot and A. Sih, 161–173. Hanover: University Press of New England.

Vanni, M. J. 1986. Competition in zooplankton communities: Suppression of small species by *Daphnia pulex*. Limnol. Oceanogr. 31:1039–1056.

Vanni, M. J. 1987a. Effects of food availability and fish predation on a zooplankton community. Ecological monographs 57:61–88.

Vanni, M. J. 1987b. Effects of nutrients and zooplankton size on the structure of a phytoplankton community. Ecology 68: 624–635.

Webster, K. and R. Peters. 1978. Some size-dependent inhibitions of larger cladocera filterers in filamentous suspensions. Limnol. Oceanogr. 23:1238–1245.

Werner, E. E. 1980. Niche theory in fisheries ecology. Trans. Amer. Fish. Soc. 109:257–260.

Werner, E. E., J. F. Gilliam, D. J. Hall and G. G. Mittelbach. 1983. An experimental test of the effects of predation risk on habitat use in fish. Ecology 64:1540–1548.

Werner, E. E. and J. F. Gilliam. 1984. The ontogenetic niche and species interactions in size-structured populations. Ann. Rev. Ecol. Syst. 15:393–425.

Wetzel, R. G. and P. H. Rich. 1973. Carbon in freshwater systems. *in*: Carbon and the Biosphere, ed. G. M. Woodwell and E. V. Pecan. U.S. AEC Symp. Ser. CONF-720510. Nat. Tech. Inform. Service, Springfield, VA.

Williamson, C. E. 1987. Predator-prey interactions between omnivorous diaptomid copepods and rotifers: The role of prey morphology and behavior. Limnol. Oceanogr. 32:167–177.

Zaret, T. M. 1980. Predation in freshwater communities. New Haven: Yale University Press.

CHAPTER 11

Size Structured Interactions in Lake Communities

Roy A. Stein, Chair, Stephen T. Threlkeld, Rapporteur, Craig D. Sandgren, W. Gary Sprules, Lennart Persson, Earl E. Werner, William E. Neill, and Stanley I. Dodson

Introduction

Historically, community ecologists have characterized organisms primarily by taxon with little regard for other potential categorizations, save perhaps for trophic position or nutrition. Certainly organisms can be classified in various ways, such as by food preferences, physiological characteristics, genotype, life history characteristics, growth rates, or size. Depending on the question to be addressed, any of these approaches could be appropriate, and a combination may perhaps be best. This chapter explores the usefulness of organism size for characterizing complex interactions in aquatic systems.

Cell size or body size of aquatic organisms is a broadly useful scaling factor for comparisons of basic metabolic processes (i.e., resource sequestering efficiency, respiration, photosynthesis rates), growth rates, production rates, reproductive commitments, and constraints on shape and mechanical design. Scaling by size also permits comparisons among ecosystems of gross structure or energy and carbon transfer efficiencies. A large literature, including several new books, attests to the importance of body size and how physiological, life history, and population parameters scale with body dimensions (Calder 1983; Peters 1983; Reynolds 1984; Schmidt-Nielsen 1984; Werner and Gilliam 1984).

The size perspective has become a useful organizing principle in the study of ecological communities just as species diversity, productivity, or trophic structure have. Size is now recognized as one of the most important attributes of an organism from both the evolutionary and ecological points of view (Werner and Gilliam 1984). Recently, at least three fundamentally different size based perspectives have developed.

As a size dependent theory about both structure and dynamics of pelagic communities, one approach emphasizes community size-spectrum analysis, including trophic groups from bacteria to fish. This approach represents the most holistic, but the least biological, perspective. In brief, Sheldon et al. (1972) observed that the living biomass in logarithmic intervals of organism size is roughly constant across the entire spectrum of plants and animals inhabiting the pelagic ocean. That is, the biomass of algae within a particular size class (logarithmic interval) is roughly the same as the biomass of algae in any other size class, which in turn is the same as the biomass of zooplankton, fish, or whales in a size class.

This remarkable observation implies that the complex ecological interactions controlling biomass concentrations in open waters balance one another quite explicitly. In fact, this regularity in ecosystem structure derives from well-established size dependent ecological and physiological relations that describe production dynamics in marine and freshwater pelagic communities (Kerr 1974; Sheldon et al. 1977; Platt and Denman 1978; Sprules et al. 1983; Sprules and Munawar 1986). Put simply, energy dissipation through a food web is balanced by decreasing turnover rates of the larger organisms occupying progressively higher trophic levels such that standing stocks are about the same in all size categories. Resource availability determines total ecosystem biomass, but energy transfer efficiencies among size-structured trophic classes determine biomass spectra. So long as trophic levels and size classes correspond, this constancy in biomass among size classes appears to hold.

This particular perspective on size structure implies that a level of order higher than the taxonomic tradition exists in aquatic communities. The dynamic implication is that change in biomass in any size interval (through predation or competition) is accompanied by compensatory changes in other size intervals such that, at some time after a perturbation, the system returns to roughly equal biomasses across the size spectrum. The waxing and waning of large *Daphnia* or colonial cyanophytes are minor spikes on the general trend of equal biomass. From this perspective, for example, one could quite legitimately speculate that the pelagic community that results in Lake Michigan when the disturbances caused by sea lamprey (*Petromyzon marinus*), alewife (*Alosa pseudoharengus*), and salmon (*Oncorhynchus spp.*) introductions approach equilibrium will be indistinguishable from the native community of the early 1900s. Certainly species will have changed, but overall size structure should be similar. Presuming no major change in nutrient loading, a particle theorist would predict that the production of 1 kg fishes (whatever species this may be) would be the same in both communities. This is a legitimate, but certainly different, view of structure.

The spectral perspective on size structure warrants careful thought for it suggests that forces grossly shaping aquatic communities are potent and knowable. Size-spectrum analysis is an exemplary tool for comparative limnology owing to its freedom from the constraints of taxonomy based structural comparisons of ecosystems as diverse as temperate lakes, tropical lakes, ancient African lakes, oceans, etc. It can be artfully used to generate hypotheses regarding comparative ecosystem functioning without initial recourse to costly organismal studies. Under specific situations of simple ecosystems, it provides a basis for remarkably precise estimates of potential production in pelagic systems (Sheldon et al. 1977) as well as insight into factors regulating higher order structure in aquatic communities that appear to be both fundamental and pervasive.

A second perspective recasting community structure in terms of organism body size appeared with the plankton studies of Hrbáček (1962) in Czechoslovakian fish ponds and similar observations by Brooks and Dodson (1965) in Connecticut lakes. From Brooks and Dodson (1965) came the size efficiency hypothesis (SEH), a hypothesis to explain the commonly observed inverse relationship between the abundances of large and small herbivorous zooplankters in freshwater lakes. Large zooplankton species are more vulnerable to planktivorous fish; however, they were hypothesized to be superior competitors in the absence of these predators, owing to their greater food collection abilities and their lower metabolic demand per unit mass. Elegantly simple in its formulation, the SEH guided plankton research into the 1980s. Most well-documented is the predation portion of the SEH in which investigators have shown that planktivorous fish can drive large-bodied zooplankton to extinction (Galbraith 1967; Wells 1970; Sprules 1972; Lynch 1979; O'Brien 1979; Vanni 1986a). Work with invertebrate predators reveals that while they do prey selectively on small-bodied species (Dodson 1974), they are unlikely to drive their prey to extinction (Hall et al. 1976; Lynch 1979). Though the predation aspect of the SEH has been confirmed, resolution of the competitive portion of the hypothesis remains elusive (Hall et al. 1976; Lynch 1977). Many studies have revealed that interaction among different-sized zooplankton rarely leads to competitive exclusion. Rather, nearly all conclude that while large herbivores can exert considerable competitive pressure on small ones, the final outcome is extremely difficult to predict (Lynch 1979, Vanni 1986b). For example, in competition experiments, Goulden et al. (1982) conclude that *Daphnia* do not consistently outcompete *Bosmina*, despite the relatively high feeding rates and low vulnerability to starvation (Threlkeld 1976) of *Daphnia*. When *Daphnia* overexploits its resources and declines in population, *Bosmina* populations expand (Goulden et al. 1982). Numerous other examples show that complicated size- and age-based interactions among zooplankton populations cause variability in the outcome of interspecific competition.

Phytoplankton studies have fostered a third size dependent approach to aquatic ecosystem studies. Margalef (1978), Lewis (1976), and Reynolds (1984) have recognized the intimate relationship between the properties of water as a fluid and the mechanical design of planktonic algae, including aspects of cell size, morphology, and coloniality. These authors have clearly demonstrated the importance of physical interactions in structuring algal assemblages. For algae, these hydrodynamic considerations, in addition to size structured grazing losses and physiological growth potentials, are of fundamental importance in dictating the species mix.

All three of these approaches have emphasized the importance of body size as a driving variable, equating adult size of a species with body size. Not until the seminal paper by Neill (1975) did ecologists begin to appreciate the fact that large species must recruit as juveniles through small size classes, and therefore are subject to the same ecological advantages and disadvantages as are the small species. In experiments in microcosms, Neill (1975) showed that a small zooplankter, *Ceriodaphnia*, could outcompete the juveniles of a larger plankter, *Daphnia*, actually driving *Daphnia* to extinction. Likewise, Reynolds and Rodgers (1983) have demonstrated a similar instance of juvenile mortality in connection with grazing on a colonial green alga. Through the demonstration of juvenile bottlenecks such as these, community ecologists have come to realize that organisms must be categorized by size as well as by species if mechanisms underlying community structure and function are to be understood. Each of these perspectives on the importance of organism size in aquatic systems has engendered significant study, experimentation, and controversy. When used judiciously with proper biological insight, mechanical design and size analysis becomes a useful additional tool in describing communities and formulating hypotheses regarding community dynamics. The following section provides a selected series of examples to demonstrate this point.

The Body Size Perspective

Evidence for Size-Structured Interactions

The size perspective has provided insight into the distribution and abundance of organisms ranging in size from bacteria to fish. For example, size dependencies of algal physiological processes help explain species' distributions across habitats. Many studies clearly document how allometric relationships govern basic phytoplankton and protozoan metabolic processes. These include potential growth rates (Fenchel 1974; Banse 1976, 1982; Findlay 1977; Malone 1980; Schlesinger et al. 1981; Reynolds 1984), nutrient uptake rates (Friebele et al. 1978; Malone 1980; Smith and Kalff 1982), the half-saturation constant for nutrient uptake (Eppley et al. 1969; Parsons and Takahashi 1973), cellular subsistence quotas for nutrients (Shuter 1978; Schlesinger et al. 1981; Smith and Kalff 1982), photosynthetic rate (Taguchi 1976; Malone 1980; Schlesinger et al. 1981; Sournia 1981), respiration rate (Laws 1975; Malone 1980), and sinking rate (Smayda 1970; Lewis 1976).

Because cell size relates to physiological and growth parameters, authors have stressed the importance of cell size in competitive interactions among phytoplankton and in seasonal or trophic succession patterns (Parsons and Takahashi 1973; Laws 1975; Margalef 1978; Smayda 1980; Turpin and Harrison 1980; Schlesinger et al. 1981; Sournia 1981; Banse 1982; Smith and Kalff 1983; Harrison and Turpin 1982; Reynolds 1984). In any phytoplankton habitat a balance must be met between the advantages usually attributed to small size (i.e., higher growth, nutrient turnover rates, and uptake efficiencies) and the disadvantages resulting from the pressure of selective grazing losses that are more severe for small cells. The optimal range of cell size must vary among habitats according to the physical, chemical, and trophic environments (Porter 1977; Shuter 1979).

Attempts have been made to relate cell size to the evolution of phytoplankton life history strategies and to environmental conditions (or successional stages) in a manner akin to the r/K selectionists' ideas generated in metazoan communities (Margalef 1978; Shuter 1979; Kilham and Kilham 1980; Harrison and Turpin 1982; Sommer 1981). Although all phytoplankton have traditionally been viewed as r-selected species relative to multicellular organisms (Kilham and Kilham 1980), the broad ranges observed in cell size, colonial complexity, and growth rates actually provide for the evolution of a gradient of physiologically and morphologically diverse life history strategies (Kilham and Kilham 1980; Sommer 1981; Reynolds 1984). In short, physiological and behavioral processes, affected by algal size, are fundamental to the distribution and abundance of individual species.

Many studies have documented the critical influence of body size in the distribution and abundance of zooplankton. Autoecological variates correlated with zooplankton size include filtering and clearance rates (MacMahon 1965; Burns 1969; DeMott 1982; Peters and Downing 1984; Knoechel and Holtby 1986a,b), threshold food levels (Stemberger and Gilbert 1985), food selection (Jacobs 1974; Gliwicz 1980; DeMott 1982), and egg production (Green 1954; Hebert 1975). These parameters have been combined with feeding patterns and rates of assimilation and respiration (also size dependent; see Peters 1983) to generate models of size-specific net energy gain (Richman 1958; Hall et al. 1976; Threlkeld 1976; Lynch et al. 1986) and optimal body size (Lynch 1977). When combined with estimates of vulnerability to predation, these models have been valuable in describing demography and distributions of zooplankton (Lynch 1977). Lynch et al. (1986) suggest that size at first maturity is the single most important property in *Daphnia* life histories. Interestingly, this parameter changes seasonally (Kerfoot 1975; Stenson 1976; Culver 1980), most likely in response to changing competitive and predatory influences.

Size structures of phytoplankton and zooplankton are to some extent interdependent. The allometric relationships described above suggest that the size structure of the grazer assemblage should influence the size structure of the phytoplankton (Carpenter and Kitchell 1984). Experimental manipulation of zooplankton size structure causes rapid changes in phytoplankton size distributions (Bergquist et al. 1985). Knoechel and Holtby (1986a) use size, not species, of zooplankton to estimate community grazing rates, arguing that "all planktonic Cladocera should possess similar filtering rates in order to coexist in a relatively homogeneous environment''. On the other hand, algal size distribution can affect the grazer size distribution. Large algal filaments inhibit feeding and growth of large daphnids (Arnold 1971; Porter 1973; Webster and Peters 1978), and may cause dominance of the zooplankton by small animals during blooms of blue-green algae (Gliwicz 1977, 1980).

The recognition of considerable within population variance in size prompted analyses of intraspecific size dependencies that complement the interspecific ones above. Among fishes, predator-prey interactions are strongly mediated by size. Piscivorous fish are size selective and gape limited (Lawrence 1957; Werner 1977), suggesting that prey for predators as diverse as esocids (*Esox spp.*, Gillen et al. 1981), largemouth bass (*Micropterus salmoides*, Lawrence 1957), yellow perch (*Perca flavescens*, Tonn and Paskowski 1986), and walleye (*Stizostedion vitreum vitreum*, Neilsen 1980) enjoy a refuge in large size. Spawning times and growth rates, as mediated through intraspecific competition, influence prey vulnerability to these predators. In work with gizzard (*Dorosoma cepedianum*) and threadfin shad (*D. petenense*), Adams and DeAngelis (1987) demonstrate how spring water temperature, through its influence on spawning times, operates to influence prey vulnerability. If rapid warming and subsequent largemouth bass spawning occur in early spring and are followed by a cooling period that delays shad spawning, then largemouth bass gain a size advantage over their shad prey. This advantage permits largemouth bass to begin feeding on young of the year shad as soon as they become available. In so doing largemouth bass maintain high growth rates, essentially matching and exceeding shad growth through summer. If young of the year largemouth bass are simply too small to feed initially on the first shad available, their growth rates on alternative foods (macroinvertebrates and zooplankton) will be so slow as to preclude feeding on young of the year shad, ultimately leading to poor overwinter survival and low recruitment (Adams and DeAngelis 1987). Similar examples exist for interactions between shad and other predators such as percids and esocids (Neilsen 1980; Carline et al. 1986).

Owing to trophic cascading, this phenomenon has important implications for community structure (Carpenter et al. 1985). During those years in which spawning times (or even prey growth rates) are such that most of the planktivore population is immune to predation, then zooplankton will be much reduced, having little or no impact on the phytoplankton (Mills et al. 1987; Mills and Forney this volume). Using data collected over 30 yr, Mills and Forney (this volume) demonstrate just these effects, with an obvious caveat. If the zooplanktivore is simply a facultative one that can use a wide variety of prey, then the impact on the zooplankton may be less predictable, with the predator switching to more profitable prey as zooplankton populations decline. This example points up the continuing value of single-species foraging studies in which an understanding of the behavioral repertoire of the predator is the primary goal. In Oneida Lake, population regulation within the perch population is also strongly size dependent. With a large cohort of perch, intense size-selective zooplanktivory reduces *D. pulex* to low levels, at which time perch switch to smaller zooplankton. Growth slows, and cannibalism by adult yellow perch and predation by walleye increase, owing to the larger window of vulnerability slow growth provides. Ultimately, this sequence of interactions leads to poor recruitment of yellow perch, not as a result of intraspecific competition for food, but rather as a result of a complex series of indirect size structured interactions (Mills et al. 1987).

Size, Predation, and Competition

Food-web interactions depend strongly on body size, and deserve closer attention. For example, many species proceed through several trophic positions during their life, but the theoretical implications of such well-known phenomena for such basic ecological interactions as competition and predation are not well defined. As aquatic organisms increase in body size, food habits and susceptibility to predators change dramatically. For example, piscivorous fishes undergo several trophic shifts, typically switching from zooplankton to macroinvertebrates to fish as they increase in size (Werner and Gilliam 1984). These shifts are associated with an increase in food particle size and usually result in enhanced growth. When prevented from shifting from zooplankton to macroinvertebrates, Eurasian perch "stunt" (Persson 1983), restricting recruitment to the piscivorous stage. Ontogenetic diet shifts also occur in a variety of other organisms including leeches, midges, damselflies, sea stars, zooplankton, and spiders (Werner and Gilliam 1984). Obviously, these changes in foraging mode vastly complicate species interactions in aquatic systems.

Via size class competition for limited resources, these niche shifts (or their absence) can have important implications for community patterns. In a particularly nice example, Hamrin and Persson (1986) explained 2-yr cycles in vendace (*Coregonus alba*) by considering size-specific intraspecific competition. In years with high recruitment, young of the year vendace compete heavily with one another as well as the soon to be reproductive 1-yr olds. Owing to poor growth of these reproductive (age 1 +) adults, they produce few young and the second summer begins with a moderate number of yearlings and few young of the year. Hamrin and Persson argue that these conditions are ideal for growth of the yearlings-initially small animals experiencing little competition from young of the year, which grow fast and have an excellent reproductive output. The next summer follows the initial pattern of strong competition between young of the year and yearlings, preventing the production of a large year class. Such patterns cannot be fruitfully explored without an appreciation of ontogenetic niche shifts that are fundamentally size dependent.

Though the sensitivity of certain life stages to ecological constraints has been appreciated for some time, the consequences for prey of size dependent bottlenecks in predator development are not well studied. Neill (this volume) discusses how survival of early instars of *Chaoborus* influence the zooplankton community in oligotrophic montane lakes, and the keystone role that *Daphnia* plays in regulating *Chaoborus* prey. Likewise, Mills and Forney (this volume) define the high summer biomass of large sized, inedible cyanobacteria in Oneida Lake as a significant bottleneck restricting energy flow to fishes. Threlkeld (1979) shows how reproductive activity of one *Daphnia* species, coupled with size-selective predation by planktivorous fish, act to regulate the density of a smaller *Daphnia* species. Although bottleneck effects are united by the presence of a critical life stage and susceptibility to some environmental constraint, the effects often ramify through the community with long-lasting effects. As a class of ecological interactions, size structured bottlenecks deserve additional study with concomitant development of unifying theory that explicitly considers size dependent processes.

Many connections among aquatic habitats involve size dependent processes (Lodge et al. this volume). In fact, competition and predation interact with habitat use by prey in a variety of ways. First, the mere presence of predators can influence the habitat distribution of their prey. Small size classes of prey suffer reduced food consumption and growth when relegated by predator intimidation to rocky substrates (Stein and Magnuson 1976; Stein 1977), littoral vegetation (Werner et al. 1983; Mittelbach 1984), or inshore areas in streams (Power 1987). When cover is limiting, a predator's presence can intensify competition by concentrating young age classes of many species into a common protective habitat (Mittelbach 1984; Mittelbach and Chesson 1987). Second, both growth rates and susceptibility to predators vary with organism size. Werner (1986) can predict habitat use and ontogenetic niche shifts of centrarchid planktivores by measuring size-specific rates of growth and predation. Ontogenetic shifts change dramatically the role of species within a community; species operate at times as competitors and at other times as predators. From this perspective, aggregation of species according to trophic levels clearly is inappropriate.

Size-specific foods and habitats influence interspecific interactions as well. By confining young of the year pumpkinseeds (*Lepomis gibbosus*) and bluegills (*L. macrochirus*) to the littoral vegetation, largemouth bass exacerbate competition between the two (Mittelbach and Chesson 1987). Indirect negative effects such as this may result in the competitive elimination of pumpkinseeds, even if the adults of these two species use separate resources. For size distributed populations, such as fishes, lizards, snails, etc., a real danger exists in judging the strength of interspecific competition from studies in which only a single stage in the life history is examined (Mittelbach and Chesson 1987).

Within homogeneous, strictly planktonic habitats, strong size dependent interactions also occur. Based on recent work by Lampert and his colleagues, zooplankton migrate vertically in response to fish predation, gaining little in terms of metabolic advantage through residence in cool hypolimnetic waters (Lampert 1987). From modeling and empirical information, Lampert (1987) concludes that fish predators, by inducing vertical migration of zooplankton grazers, can indirectly enhance algal production, favoring small, fast growing edible forms. Clearly, top-down effects instigated by predaceous fishes are important, acting not just to release algal communities from direct grazing pressure, but also serving to shift the entire species/size community of phytoplankton. These patterns clearly have interesting implications for nutrient cycling and system productivity (Shapiro and Wright 1984). So too do the vertical migrations of flagellate algae (Heaney and Talling 1980; Sandgren in press), a phenomenon that is generally restricted to large celled forms because of the necessity of overcoming epilimnetic turbulence to effect directional movement. At an even finer scale of spatial/habitat resolution, the occurrence of micropatches of nutrients released by zooplankton may influence coexistence of algal species whose ability to process these patches depends largely on cell-size dependent nutrient uptake kinetics (Lehman and Scavia 1982, Reinertson et al. 1986). And, the occurrence of size/age structured swarms of zooplankton and perhaps fish in the open pelagial may influence a whole range of species interactions in the epilimnion (Folt 1987).

Many other sorts of interactions in aquatic food webs are size dependent. Because many invertebrate predators are size limited, increasing effective prey size (in response to predator presence) reduces prey vulnerability. In a recent review, Havel (1987) notes that protozoans, rotifers, cladocerans, and bryozoans can modify their morphology (usually by building spines or changing shape) in response to chemical cues produced by prey or predator, and thus reduce their vulnerability to invertebrate predators. Other types of chemically related size dependencies could include both the effects of size on the quantity of exudate released and the volume of water over which a given signal is dispersed (via swimming speed, surface area effects, or metabolic rate). These responses serve to complicate already complex size structured interactions and should be pursued in future research.

Other Size Dependent Processes

As the previous examples suggest, body size usually manifests itself through predation or competition; however, other size related phenomena exist. For example, overwinter mortality in age-O brook trout (*Salvelinus fontinalis*, Hunt 1969), smallmouth bass (*Micropterus dolomieui*, Shuter et al. 1980), largemouth bass (Adams et al. 1982), and tiger muskellunge (*Esox* hybrid, Carline et al. 1986) depends upon fish length at the end of autumn. Small fish survive less well than large ones. In several attempts to formulate life history strategies for planktonic algae based upon physiological competency to survive in different types of physical/chemical planktonic habitats, investigators have ultimately resorted to algal size as a primary scaling factor, as discussed above. An invading crayfish, *Orconectes rusticus*, is larger than the native species it typically displaces, *O. virilis*, *O. propinquus*, or *O. sanborni* (Capelli 1982; Butler and Stein 1985). One mechanism suggested for the rapid replacement by *O. rusticus* is that large males sequester females of both species, dramatically reducing the reproductive success of the native species (Butler and Stein 1985). These few examples emphasize the pervasive effects body size can have, from directly influencing survival to controlling individual reproductive success.

Implications of a Size Structured Perspective

As a particularly useful analytical approach, the body size perspective has influenced both our ecological and evolutionary concept of aquatic community function. But whereas body size dependencies have suggested processes or interactions to investigate (size-specific food habits, roles as competitors or predators, etc.), the logical link between evolutionary theory (which is age-dependent) and size dependent ecological theory has been especially difficult to forge. The biological and, hence, evolutionary basis of community structure must not be forgotten.

Whereas population models without age or size structure provide changes in total numbers (i.e., dn/dt), a basic continuous time model of a size structured population provides changes in the size distribution (Werner and Gilliam 1984). By explicitly including individual growth rates into a balance or continuity equation (Nisbet and Gurney 1983), Werner and Gilliam (1984) were able to model the rate of increase, r, of a population with a stable size distribution rather than age distribution. Based on their review, little has been done on density dependence of size structured populations, with few studies explicitly dealing with two interacting, size-structured populations (Werner and Gilliam 1984). Clearly, more work is necessary in these areas.

Given the importance of size structured interactions, experimental protocols should account for size structured effects. If an entire size structured population is added or removed, it will probably not be clear which size classes are primarily responsible for observed effects. Conversely, if the experimental unit (pond, bag, beaker) fails to allow the full range of normal ecological constraints on the size classes present, then the experimental results will more strongly reflect that limitation of the experimental system than the natural range of effects of the species. Experimental systems are generally abstractions of natural complex systems. Typical of this approach is the use of arctic and alpine systems, where trophic levels may be represented by a single species. Whereas results from these systems may indicate the range of responses possible when entire trophic levels are manipulated, these experiments do not provide a basis for inferring how a single species may react in a more complex system. Removal of one species may simply be compensated for by the response of another formerly rare species. In our experience, these relatively simple systems have provided the clearest evidence of top-down control.

Experimentalists and modelers must carefully consider the tremendous differences in time scales at which interacting organisms in aquatic communities operate (Frost et al. this volume). Generation times show strong allometric relationships and range from minutes for bacteria to years for fishes and clonal macrophytes, with other important life histories falling between these extremes. The time scales employed by experiments and models may determine their applicability to community dynamics in whole lakes.

Comparable trophic-level manipulations consisting of removal or addition of a predator species are often difficult to achieve at lower trophic levels.For instance, adding nutrients usually favors only a subset of the available algal species pool, and is thus comparable to elimination or addition of a single species of fish while the remainder of the fish community is unaltered. Manipulating piscivores in this fashion might result only in the transfer of planktivore control to a different piscivorous species. Similarly, if the experimental system includes only one member of a suite of species of similar size or action, then the observed results are specific perhaps only to the artificial community and not to the natural community being abstracted. Due to these design constraints, we do not know how size structured interactions at lower trophic levels influence other trophic levels.

Trade-offs emanate from these considerations. First, though natural systems may provide the only setting in which all ecological constraints are represented, and fundamental structural patterns are manifested, statistical resolution of any response within nature is obviously tenuous (Diamond 1984; Hurlbert 1984). In contrast, small experimental systems (bags, tanks, ponds) lend insight (biological and statistical) into a reasonable subset of the natural interactions. Designs of these experiments must incorporate manipulations to account for potential interactions across several trophic levels and a broad size range of organisms. To accommodate these needs, experimental units must increase in size; yet, with this increase comes a commensurate reduction in number of replicates, as demonstrated in a recent review of articles published in *Ecology* (Karieva pers. comm.). Reducing the number of replicates conflicts directly with recent, widely accepted advice given by Hurlbert (1984). To resolve this conflict, experimental community ecologists working in aquatic systems with size dependent processes must explicitly consider the spatial dimension in their experiments, being forced perhaps into doing experiments across a range of plot sizes, including bays, large enclosures, limnocorrals, and partial and whole lakes.

Though we have argued for aggregating species into size classes, this technique may obscure important relationships. We believe that neither taxon-specific nor size-specific approaches are in themselves adequate. For example, size has not always been successfully incorporated into the ecological framework; these exceptions lend insight into ordering principles in ecological communities. For example, basic properties of algal cells such as POC/cell or maximum potential growth rate are demonstrably size dependent among planktonic species of diatoms, green algae, chryso-phytes, and dinoflagellates, but the *slopes* of the relationships differ for some of these phylogenetically remote groups, as do the absolute values of the parameters for most cell size classes (Kennedy 1984; Tilman et al. 1986). Such physiological differences in combination with size selective grazing could well be important in influencing the distribution and relative dominance of algal species, influencing turnover rates of nutrients and carbon within the ecosystem. Analysis of both size spectra and the or-ganismal composition of at least the lower trophic levels must occur for accurate ecosystem comparisons, and the physiological and reproductive potential of the dominant species must be well known. In this respect, the size spectra analyses are valuable in generating testable hypotheses regarding basic comparative organismal physiology.

Combinations of various aggregating strategies can generate important insights and hypotheses when coupled with adequate basic understanding of the behavior, life history, and metabolic and reproductive potential of the species involved. Sprules (1984) attempted to describe community composition with different aggregation strategies (size, functional groups, species, etc.) and found that a combination of aggregation types was most helpful. Obvious as well is that the strategy chosen should be geared to the question being asked. Drenner et al. (1986) used several methods of measuring algal responses to planktivorous fish; clearly, trade-offs existed between time and cost of analysis and the qualities of the insights derived from each kind of effort. In light of the significant advances in ecological understanding that have resulted from the use of organism size in place of, or in addition to, the more traditional taxon approach to aggregation, the development and application of other bases for species aggregation (i.e., life history strategies, physiological characteristics, etc.) should be encouraged, and a focused study of the most robust way to analyze size structured interactions would be most helpful.

Pragmatic Concerns

To accommodate intensive study of size structured interactions, some modifications to our present funding structure may be appropriate. An understanding of community level processes require, by definition, ex-pertise across trophic levels. To improve upon our empirical science, a record that often has included inappropriate lumping strategies and ma-nipulation experiments, we require efforts that incorporate expertise spanning the body size continuum from the microbial loop to fish. Studying these systems requires not only this pragmatic expertise but also the theoretical underpinnings that only collaboration with a theorist can bring. In our chapter, we identified several areas where a theoretical perspective on size structured interactions would be helpful. For example, are conditions for coexistence of two species similar to those predicted by traditional competition theory, in the face of this myriad of differentially functioning size classes? Questions such as this one could be profitably explored in aquatic systems.

Operationally, we can infuse expertise into community ecology in several ways. First, single individuals working either in population ecology/autecology and theory can contribute dramatically to our progress. Understanding mechanistic interactions among individuals within a size structured system can provide much needed detailed understanding of higher level processes, if these studies are well couched in the conceptual framework of community ecology. In turn, theorists working on size structured interactions should be able to provide much insight that ultimately can guide and direct community level questions. Individual investigators contributing basic ecological, physiological, and reproductive information about aquatic organisms is still a vital concern in aquatic ecology. Our perceptions about the potentials of phytoplankton and zooplankton species, especially, is based upon too few, well-studied example organisms. Particularly, the relative importance of phylogenetic history vs simple gross size in dictating the physiological and reproductive characteristics of these organisms requires continued investigation.

While these options are practical (single investigators, low budgets), we believe that other avenues for progress exist as well. For certain problems, collaborative multi-investigator efforts may be the best approach. Through multiple empiricist/theorist interactions, we believe a good deal of insight into size structured, complex interactions can be gained.

Owing to the complexity of these interactions, complete, fully replicated experiments across time and space scales will be necessary. As such, facilities that range from wet laboratories (including facilities for replicate chemostats or microcosms) to fiberglass pools to ponds and lakes will be required. Where possible, these facilities should be shared (consistent with the call for multi-investigator collaborative efforts) and the opportunity to collaborate with state/federal agencies, where hatchery-pond facilities are available, should not be missed. Commensurate with the inclusion of multiple size classes per species within experimental protocol comes the necessity for large numbers of replicate systems. Large manipulative efforts will be necessary if we are to begin to understand the complexity of size structured interactions in aquatic ecosystems.

References

Adams, S. M., and D. L. DeAngelis. 1987. Indirect effects of bass-shad interactions on predator population structure and food web dynamics. *in*: Predation: direct and indirect impacts on aquatic communities, ed. W. C. Kerfoot and A. Sih, 103–117. Hanover: University Press of New England.

Adams, S. M., R. B. McLean, and M. M. Huffman. 1982. Structuring of a predator population through temperature-mediated effects on prey availability. Can. J. Fish. Aquat. Sci. 39:1175–1184.

Arnold, J. D. 1971. Ingestion, assimilation, survival and reproduction by *Daphnia pulex* fed seven species of blue-green algae. Limnol. Oceanogr. 16:906–920.

Banse, K. 1976. Rates of growth, respiration and photosynthesis of unicellular algae as related to cell size—a review. J. Phycol. 12:135–140.

Banse, K. 1982. Cell volumes, maximal growth rates of unicellular algae and ciliates, and the role of ciliates in the marine pelagial. Limnol. Oceanogr. 27:1059–1071.

Bergquist, A. M., S. R. Carpenter, and J. C. Latino. 1985. Shifts in phytoplankton size structure and community composition during grazing by contrasting zooplankton assemblages. Limnol. Oceanogr. 30:1037–1045.

Brooks, J. L., and S. I. Dodson. 1965. Predation, body size, and composition of plankton. Science 150:28–35.

Burns, C. W. 1969. Relation between filtering rate, temperature, and body size in four species of *Daphnia*. Limnol. Oceanogr. 14:693–700.

Burns, C. W. and J. J. Gilbert. 1986. Effects of daphnid size and density on interference between *Daphnia* and *Keratella cochlearis*. Limnol. Oceanogr. 31:848–858.

Butler, M. J., IV and R. A. Stein. 1985. An analysis of the mechanisms governing species replacements in crayfish. Oecologia 66:168–177.

Calder, W. A. III. 1983. Ecological scaling: mammals and birds. Ann. Rev. Ecol. Syst. 14:213–230.

Carpenter, S. R. and J. F. Kitchell. 1984. Plankton community structure and limnetic primary production. Am. Nat. 124: 159–172.

Carpenter, S. R., J. F. Kitchell, and J. R. Hodgson. 1985. Cascading trophic interactions and lake productivity. BioScience 35:634–639.

Capelli, G. M. 1982. Displacement of native crayfish by *Orconectes rusticus* in northern Wisconsin. Limnol. Oceanogr. 27:741–745.

Carline, R. F., R. A. Stein, and L. M. Riley. 1986. Effects of size at stocking, season, largemouth bass predation, and forage abundance on survival of tiger muskellunge. Am. Fish. Soc. Spec. Publ. 15:151–167.

Culver, D. A. 1980. Seasonal variation in the sizes at birth and first reproduction in Cladocera. Am. Soc. Limnol. Oceanogr. Spec. Symp. 3:358–366.

DeMott, W. R. 1982. Feeding selectivities and relative ingestion rates of *Daphnia* and *Bosmina*. Limnol. Oceanogr. 27:518–527.

Diamond, J. 1984. Overview: laboratory experiments, field experiments, and natural experiments. *in*: Community ecology, ed. J. Diamond and T. J. Case, 3–22. New York: Harper and Row.

Dodson, S. I. 1974. Zooplankton competition and predation: an experimental test of the size-efficiency hypothesis. Ecology 55:605–613.

Drenner, R. W., S. T. Threlkeld, and M. P. McCracken. 1986. Experimental analysis of the direct and indirect effects of filter feeding clupeids on plankton community structure. Can. J. Fish. Aquat. Sci. 43:1935–1945.

Eppley, R., J. N. Rogers, and J. J. McCarthy. 1969. Half-saturation constants for uptake of nitrate and ammonium by marine phytoplankton. Limnol. Oceanogr. 14:912–920.

Fenchel, T. 1974. Intrinsic rate of natural increase: the relationship with body size. Oecologia 14:317–326.

Finlay, B. J. 1977. The dependence of reproductive rate on cell size and temperature in freshwater ciliated protozoa. Oecologia 30:75–81.

Folt, C. L. 1987. An experimental analysis of costs and benefits of zooplankton aggregation. in: Predation: direct and indirect impacts on aquatic communities, ed. W. C. Kerfoot and A. Sih, 300–314. Hanover: University Press of New England.

Friebele, S., D. L. Correll, and M. A. Faust. 1978. Relationship between phytoplankton cell size and the rate of orthophosphate uptake: in situ observations of an estuarine population. Mar. Biol. 45:39–52.

Galbraith, M. G. 1967. Size-selective predation on *Daphnia* by rainbow trout and yellow perch. Trans. Am. Fish. Soc. 96: 1–10.

Gillen, A. L., R. A. Stein, and R. F. Carline. 1981. Predation by pellet-reared tiger muskellunge on minnows and bluegills in experimental systems. Trans. Am. Fish. Soc. 110:197–209.

Gliwicz, Z. M. 1977. Food size selection and seasonal succession of filter-feeding zooplankton in an eutrophic lake. Ekol. Polska 25:179–225.

Gliwicz, Z. M. 1980. Filtering rates, food size selection, and feeding rates in cladocerans: another aspect of interspecific competition in filter-feeding zooplankton. Am. Soc. Limnol. Oceanogr. Spec. Symp. 3:282–291.

Goulden, C. E., L. L. Henry, and A. J. Tessier. 1982. Body size, energy reserves, and competitive ability in three species of Cladocera. Ecology 63:1780–1789.

Green, J. 1954. Size and reproduction in *Daphnia magna*. Proc. Zool. Soc. London 124:535–545.

Hall, D. J., S. T. Threlkeld, C. W. Burns, and P. H. Crowley. 1976. The size-efficiency hypothesis and the size structure of zooplankton communities. Ann. Rev. Ecol. Syst. 7:177–208.

Hamrin, S. F., and L. Persson. 1986. Asymmetrical competition between age classes as a factor causing population oscillations in an obligate planktivorous fish species. Oikos 47:223–232.

Harrison, P. J., and D. H. Turpin. 1982. The manipulation of physical, chemical, and biological factors to select species from natural phytoplankton communities. in: Marine mesocosms: biological and chemical research in experimental ecosystems, ed. G. Grice and M. Reeve, 275–291. New York: Springer-Verlag.

Havel, J. E. 1987. Predator-induced defense: a review. in: Predation: direct and indirect impacts on aquatic communities, ed. W. C. Kerfoot and A. Sih, 263–278. Hanover: University Press of New England.

Heaney, S. I., and J. F. Talling. 1980. Dynamic aspects of dinoflagellate distribution patterns in a small, productive lake. J. Ecol. 68:75–94.

Hebert, P. D. N. 1975. Enzyme variability in natural populations of *Daphnia magna*. I. Population structure in East Anglia. Evolution 28:546–556.

Hrbáček, J. 1962. Species composition and the amount of zooplankton in relation to fish stock. Rozpr. Cesk. Akad. Ved Rada Mat. Prir. Ved 72:1–116.

Hunt, R. L. 1969. Overwinter survival of wild fingerling brook trout in Lawrence Creek, Wisconsin. J. Fish. Res. Board Can. 26:1473–1483.

Hurlbert, S. H. 1984. Pseudoreplication and the design of ecological field experiments. Ecol. Monogr. 54:187–211.

Kennedy, K. C. 1984. The influence of cell size and the colonial habit on phytoplankton growth and nutrient uptake kinetics: a critical evaluation using a phylogenetically related series of volvocene green algae. MS thesis. Univ. Texas-Arlington.

Jacobs, J. 1974. Quantitative measurement of food selection. Oecologia 14: 413–417.

Kerfoot, W. C. 1975. The divergence of adjacent populations. Ecology 56:1298–1313.

Kerr, S. R. 1974. Theory of size distributions in ecological communities. J. Fish. Res. Board Can. 31:1859–1862.

Kilham, P., and S. S. Kilham. 1980. The evolutionary ecology of phytoplankton. *in*: The physiological ecology of phytoplankton, ed. I. Morris, 571–597. Berkeley: University of California Press.

Knoechel, R. and L. B. Holtby. 1986a. Construction and validation of a body-length-based model for the prediction of cladoceran community filtering rates. Limnol. Oceanogr. 31:1–16.

Knoechel, R. and L. B. Holtby. 1986b. Cladoceran filtering rate: body length relationships for bacterial and larger algal particles. Limnol. Oceanogr. 31:195–200.

Lampert, W. 1987. Vertical migration of freshwater zooplankton: indirect effects of vertebrate predators on algal communities. *in*: Predation: direct and indirect impacts on aquatic communities, ed. W. C. Kerfoot and A. Sih, 291–299. Hanover: University Press of New England.

Lawrence, J. M. 1957. Estimated sizes of various forage fishes largemouth bass can swallow. Proc. Ann. Conf. Southeast. Assoc. Game Fish Comm. 11:220–225.

Laws, E. A. 1975. The importance of respiration losses in controlling the size distributions of marine phytoplankton. Ecology 56:419–426.

Lehman, J. T., and D. Scavia. 1982. Microscale patchiness of nutrients in plankton communities. Science 216:729–730.

Lewis, W. M., Jr. 1976. Surface/volume ratio: implications for phytoplankton morphology. Science 192:885–887.

Lynch, M. 1977. Fitness and optimal body size in zooplankton populations. Ecology 58:763–774.

Lynch, M. 1979. Predation, competition, and zooplankton community structure: an experimental study. Limnol. Oceanogr. 24:253–274.

Lynch, M., L. J. Weider, and W. Lampert. 1986. Measurement of the carbon balance in *Daphnia*. Limnol. Oceanogr. 31: 17–33.

Malone, T. C. 1980. Algal size. *in*: The physiological ecology of phytoplankton, ed. I. Morris, 433–463. Berkeley: University of California Press.

Margalef, R. 1978. Life-forms of phytoplankton as survival alternatives in an unstable environment. Oceanol. Acta 1: 493–509.

McMahon, J. W. 1965. Some physical factors influencing the feeding behavior of *Daphnia magna* Straus. Can. J. Zool. 43:603–612.

Mills, E. L., J. L. Forney, and K. J. Wagner. 1987. Fish predation and its cascading effect on the Oneida Lake food chain. *in*: Predation: direct and indirect impacts on aquatic communities, ed. W. C. Kerfoot and A. Sih, 118–131. Hanover: University Press of New England.

Mittelbach, G. G. 1984. Predation and resource partitioning in two sunfishes (Centrarchidae). Ecology 65:499–513.

Mittelbach, G. G., and P. L. Chesson. 1987. Predation risk: indirect effects on fish populations. *in*: Predation: direct and indirect impacts on aquatic communities, ed. W. C. Kerfoot and A. Sih, 315–332. Hanover: University Press of New England.

Neill, W. E. 1975. Experimental studies of microcrustacean competition, community composition and efficiency of resource utilization. Ecology 56:809–826.

Neilsen, L. A. 1980. Effect of walleye (*Stizostedion vitreum vitreum*) predation on juvenile mortality and recruitment of yellow perch (*Perca flavescens*) in Oneida Lake, New York. Can. J. Fish. Aquat. Sci. 37:11–19.

Nisbet, R. M., and W. S. C. Gurney. 1983. The systematic formulation of population models for insects with dynamically varying in-star duration. Theor. Popul. Biol. 23:114–135.

O'Brien, W. J. 1979. The predator-prey interaction of planktivorous fish and zooplankton. Am. Sci. 67:572–581.

Parsons, T. R., and M. Takahashi. 1973. Environmental control of phytoplankton cell size. Limnol. Oceanogr. 18:511–515.

Persson, L. 1983. Food consumption and competition between age classes in a perch (*Perca fluviatilis*) population in a shallow eutrophic lake. Oikos 40:197–207.

Peters, R. H. 1983. The ecological implications of body size. New York: Cambridge University Press.

Peters, R. H. and J. A. Downing. 1984. Empirical analysis of zooplankton filtering and feeding rates. Limnol. Oceanogr. 29:763–784.

Platt, T., and K. Denman. 1978. The structure of pelagic marine ecosystems. Rapp. P.-V. Reun. Const. Int. Explor. Mer 173: 60–65.

Porter, K. G. 1973. Selective grazing and differential digestion of algae by zooplankton. Nature 244:179–180.

Porter, K. G. 1977. The plant-animal interface in freshwater ecosystems. Am. Sci. 65:159–170.

Power, M. E. 1987. Predator avoidance by grazing fishes in temperate and tropical streams: importance of stream depth and prey size. *in*: Predation: direct and indirect impacts on aquatic communities, ed. W. C. Kerfoot and A. Sih, 333–351. Hanover: University Press of New England.

Reinertsen, H., A. Jenson, A. Langeland, and Y. Olsen. 1986. Algal competition for phosphorus: the influence of zooplankton and fish. Can. J. Fish. Aquat. Sci. 43:1135–1141.

Reynolds, C. S. 1984. The ecology of freshwater phytoplankton. Cambridge: Cambridge University Press.

Reynolds, C. S., and M. W. Rodgers. 1983. Cell- and colony-division in *Eudorina* (Chlorophyta: Volvocales) and some ecological implications. Br. Phycol. J. 18:111–119.

Richman, S. 1958. The transformation of energy by *Daphnia pulex*. Ecol. Monogr. 28:273–291.

Sandgren, C. D. in press. The ecology of chrysophyte flagellates: their growth and perennation strategies as freshwater phytoplankton. *in*: Growth and reproductive strategies of freshwater phytoplankton, ed. C. D. Sandgren. Cambridge: Cambridge University Press.

Schlesinger, D. A., L. A. Molot, and B. J. Shuter. 1981. Specific growth rates of freshwater algae in relation to cell size and light intensity. Can. J. Fish. Aquat. Sci. 38:1052–1058.

Schmidt-Neilsen, K. 1984. Scaling, why is animal size so important? Cambridge: Cambridge University Press.

Shapiro, J., and D. I. Wright. 1984. Lake restoration by biomanipulation. Round Lake, Minnesota, the first two years. Freshwat. Biol. 14:371–383.

Sheldon, R. W., A. Prakash, and W. H. Sutcliffe. 1972. The size distribution of particles in the ocean. Limnol. Oceanogr. 17:329–339.

Sheldon, R. W., W. H. Sutcliffe, and M. A. Paranjape. 1977. Structure of pelagic food chains and relationship between plankton and fish production. J. Fish. Res. Board Can. 34: 2344–2353.

Shuter, B. J. 1978. Size dependence of phosphorus and nitrogen subsistence quotas in unicellular microorganisms. Limnol. Oceanogr. 23:1248–1255.

Shuter, B. J. 1979. A model of physiological adaptation in unicellular algae. J. Theor. Biol. 78:519–552.

Shuter, B. J., J. A. MacLean, F. E. J. Fry, and H. A. Regier. 1980. Stochastic simulation of temperature effects on first-year survival of smallmouth bass. Trans. Am. Fish. Soc. 109:1–34.

Smayda, T. J. 1970. The suspension and sinking of phytoplankton in the sea. Oceanogr. Mar. Biol. Ann. Rev. 8:353–414.

Smayda, T. J. 1980. Phytoplankton species succession. in: The physiological ecology of phytoplankton, ed. I. Morris, 493–571. Berkeley: University of California Press.

Smith, R. E. H., and J. Kalff. 1982. Size dependent phosphorus uptake kinetics and cell quota in phytoplankton. J. Phycol. 18:275–284.

Smith, R. E. H. and J. Kalff. 1983. Competition for phosphorus among co-occurring freshwater phytoplankton. Limnol. Oceanogr. 28:448–464.

Sommer, U. 1981. The role of r- and K- selection in the succession of phytoplankton in Lake Constance. Acta Oecol. Gen. 2:327–342.

Sournia, A. 1981. Morphological bases of competition and succession. Can. Bull. Fish. Aquat. Sci. 210:339–346.

Sprules, W. G. 1972. Effects of size-selective predation and food competition on high-altitude zooplankton communities. Ecology 53:375–386.

Sprules, W. G. 1984. Towards an optimal classification of lake zooplankton for lake ecosystem studies. Verh. Int. Verein. Limnol. 22:320–325.

Sprules, W. G., J. M. Casselman, and B. J. Shuter. 1983. Size distributions of pelagic particles in lakes. Can. J. Fish. Aquat. Sci. 40:1761–1765.

Sprules, W. G., and M. Munawar. 1986. Plankton size spectra in relation to ecosystem productivity, size, and perturbation. Can. J. Fish. Aquat. Sci. 43:1789–1794.

Stein, R. A. 1977. Selective predation, optimal foraging, and the predator-prey interaction between fish and crayfish. Ecology 58:1237–1253.

Stein, R. A., and J. J. Magnuson. 1976. Behavioral response of crayfish to a fish predator. Ecology 57:751–761.

Stemberger, R. S. and J. J. Gilbert. 1985. Body size, food concentration, and population growth in planktonic rotifers. Ecology 66:1151–1159.

Stenson, J. A. E. 1976. Significance of predator influence on compositon of *Bosmina* spp. populations. Limnol. Oceanogr. 21:814–822.

Taguchi, S. 1976. Relationship between photosynthesis and cell size of marine diatoms. J. Phycol. 12:185–189.

Threlkeld, S. T. 1976. Starvation and the size structure of zooplankton communities. Freshwat. Biol. 6:489–497.

Threlkeld, S. T. 1979. The midsummer dynamics of two *Daphnia* species in Wintergreen Lake, Michigan. Ecology 60:165–179.

Tilman, D., R. Kiesling, R. Sterner, S. S. Kilham, and F. A. Johnson. 1986. Green, bluegreen and diatom algae: taxonomic differences in competitive ability for phosphorus, silicon and nitrogen. Arch. Hydrobiol. 106:473–485.

Tonn, W. M. and C. A. Paszkowski. 1986. Size-limited predation, winter-kill, and the organization of *Umbra-Perca* fish assemblages. Can. J. Fish. Aquat. Sci. 43:194–202.

Turpin, D. L. and P. J. Harrison. 1980. Cell size manipulation in natural marine, planktonic, diatom communities. Can. J. Fish. Aquat. Sci. 37:1193–1195.

Vanni, M. J. 1986a. Fish predation and zooplankton demography: indirect effects. Ecology 67:337–354.

Vanni, M. J. 1986b. Competition in zooplankton communities: suppression of small species by *Daphnia pulex*. Limnol. Oceanogr. 31:1039–1056.

Webster, K. E. and R. H. Peters. 1978. Some size-dependent inhibitions of larger cladoceran filtering in filamentous suspensions. Limnol. Oceanogr. 23:1238–1245.

Wells, L. 1970. Effects of alewife predation on zooplankton populations in Lake Michigan. Limnol. Oceanogr. 15: 556–565.

Werner, E. E. 1977. Species packing and niche complementarity in three sunfishes. Am. Nat. 111:553–578.

Werner, E. E. 1986. Species interactions in freshwater fish communities. *in*: Community ecology, ed. J. Diamond and T. J. Case, 344–358. New York: Harper and Row.

Werner, E. E., and J. F. Gilliam. 1984. The ontogenetic niche and species interactions in size-structured populations. Ann. Rev. Ecol. Syst. 15:393–425.

Werner, E. E., J. F. Gilliam, D. J. Hall, and G. G. Mittelbach. 1983. An experimental test of the effects of predation risk on habitat use in fish. Ecology 64:1540–1548.

CHAPTER 12

Spatial Heterogeneity and Habitat Interactions in Lake Communities

David M. Lodge, Rapporteur, John W. Barko, Chair,
David Strayer, John M. Melack, Gary G. Mittelbach,
Robert W. Howarth, Bruce Menge, and John E. Titus

Introduction

Much of our understanding of complex interactions and community structure in lakes has come from studies of planktonic communities. One has only to look at the ideas and examples described in this volume to see how successful these studies have been, both at elucidating the workings of planktonic communities and at contributing concepts to the broader field of ecology. In our opinion, this success can be attributed to the relative ease of sampling and manipulating planktonic communities, the relative simplicity of these communities, and to the long tradition of talented ecologists who have worked on this community.

Despite the unassailable history of productivity in pelagic zone ecology and the many current questions that will ensure continued productivity, it is clear that any singular emphasis on the pelagic zone will limit the ultimate contributions of limnologists to community ecology. If limnologists are to maximize their contributions to community ecology, the littoral zone and profundal benthos must also be included in our research agenda.

There are at least three reasons for lacustrine ecologists to extend their gaze beyond the pelagic zone. First, the pelagic zone is not an isolated ecosystem. As we will emphasize below, there are numerous links among pelagic communities, littoral communities, and profundal communities. Some of these links clearly exert strong effects on the pelagic communities of most lakes, while others are so weak that they can be safely ignored in all but the most extraordinary circumstances. These links must be recognized and understood if we are to achieve a general understanding of pelagic communities.

Second, comparative studies conducted in a variety of habitats are helpful in revealing aspects of community dynamics that are not apparent from studies confined to a single habitat. This may be due to the structure of the habitat itself, which may impose unrecognized constraints on the community, or to the composition of the species pool, which may limit the dynamics of the community in an unappreciated way. For example, the pelagic zone has less patchiness and fewer refuges from visual predators than other environments (cf. Folt 1987). Its biological communities are also relatively species poor. These features probably influence the structure and dynamics of the pelagic community, but cannot be fully appreciated without reference to work done in other, more complex habitats.

Finally, there are many important research questions about communities in the littoral and profundal zones that remain unanswered. However, few of the other chapters in this volume address non-pelagic habitats. Therefore, the first section of this chapter, "Complex Interactions in the Benthos," introduces a few central topics for research within the littoral and profundal zones. (Throughout this chapter, we use the term benthos to include both littoral and profundal habitats.) While this list of topics for benthic research is certainly not exhaustive, it highlights the extent to which important questions about the lacustrine benthos have been neglected.

In the second section of this chapter, "Benthic-Pelagic Links," we proceed to examine some of the important interactions between benthic and pelagic habitats. In most cases, the links have to do with the movements of animals across habitat boundaries. The currency of other links, however, is dissolved nutrients. These links emphasize that biogeochemical and trophic considerations are inextricably bound.

In the third section of this chapter, "Factors Affecting the Strength of Interhabitat Links," we examine four major factors that affect the strength of interhabitat links. In this and the previous two sections, we rely on current literature to guide our discussions. However, we do not hesitate to speculate on topics that may have extreme importance, but which limnologists have not yet explored.

In our final section, "Research Needs and New Experimental Approaches," we outline a few major obstacles and partial solutions—both scientific and institutional—to approaching the questions emphasized earlier in the chapter.

This chapter is not a review of the state of our knowledge of benthic ecology or of interhabitat links. It is selective, not comprehensive. It also should not be taken as a clear-cut agenda for research. Rather, its eclectic and speculative nature is meant to stimulate thought in new areas. Because we have written about questions to which there are no clear answers, only further research will enable us to identify the useful ideas.

Complex Interactions in the Benthos

The Microbial Component in Sediments

Porter et al. (this volume) focus considerable attention on the nature and significance of the microbial component in the pelagic zone of lakes. Two key points that emerge are that much of planktonic primary production may flow through this component and that the small size of microbes makes much of this production relatively unavailable to large zooplankton. Thus, if microbial production is to reach fish, it must pass through intermediate consumers, with a concomitant loss in efficiency. Similarly, large consumers are unlikely to exert strong, direct control over the microbial community.

We expect a very different situation in the sediments, where we postulate that microbial production is directly available to large consumers. We make this assertion for two reasons. First, much of the benthic microbial community is attached to sediment particles. It is mechanically simpler for a large (ca 1-10 mm) consumer to ingest a sediment particle of dimension 10-1000 μm than to capture a free bacterial cell less than 1 μm long. Second, concentrations of interstitial bacteria may be very high (10^2-10^3 times higher than concentrations in the open water), so a filter feeder that could not capture enough bacteria to survive in the pelagic zone could survive on interstitial water. One example is the small bivalve *Pisidium*, which feeds on interstitial bacteria in lake sediments (Lopez and Holopainen 1987).

The postulated availability of benthic microbes to macroinvertebrates has at least two important consequences. First, bacterial production can be transferred to fish through a single intermediate consumer (e.g., the midge *Chironomus*), at relatively high overall efficiency. Second, in possible contrast to the situation in the pelagic zone, metazoan consumers may exert effective direct control over the microbial component in the sediments.

Is There an Independent Meiofaunal Food Web in Lakes?

The meiofauna (benthic metazoans too small to be retained on a 0.5 mm screen) has received much attention from marine biologists, but has been largely neglected by limnologists. One interesting suggestion from marine research is that the meiofauna and macrofauna are only loosely coupled. Two observations support this suggestion. The size structure of marine benthic communities is bimodal, with many large animals (10^{-4} to 10^0 g dry wt.) and many small animals (10^{-8} to 10^{-6} g dry wt.), but few animals of intermediate size (Warwick 1984). Second, observations of trophic interactions show that the macrofaunal and meiofaunal food webs are only loosely connected (McIntyre 1969; Reise 1979). Thus, the marine meiofauna and macrofauna appear to operate in parallel, rather than in series, in benthic food webs. The meiofauna may therefore compete for carbon with the macrofauna. This scenario is in some ways analogous to the microbial component of the pelagic zone.

As scanty as evidence supporting the idea of an independent meiofaunal food web is for marine sediments, our knowledge of the freshwater meiofauna is even scantier. The size structure of the zoobenthos is known for only one lake, where it is strikingly different from those of marine sites (Strayer 1986). In particular, the bimodality characteristic of marine size spectra is entirely lacking, and there is no evidence of a trophic disjunction between the freshwater meiofauna and macrofauna (Strayer 1968). Is there a meiofaunal food web in lakes that is largely unconnected to the macrofaunal food web? Are there fundamental differences in the structure and function of the lacustrine and marine benthos? These are answerable questions that limnologists have not yet investigated.

Omnivory, Predation, and Body Size in the Benthos

Sediment-associated benthic animals are faced with an assortment of small particles with a wide range of nutritional values (Fig. 12.1). We might therefore expect small animals to have more specialized, higher quality diets than large animals, which must necessarily ingest a certain amount of sediment to acquire algal and bacterial particles, which are of higher quality. The predictions of this model are far-reaching:

1. Some small benthic animals may be dietary specialists, while larger animals may have broad diets and exhibit omnivory.
2. Assimilation efficiencies should rise with decreasing body size.
3. Ingestion/production should therefore decline with decreasing body size.
4. Although macrofaunal species apparently are unable to use microbial biomass as their chief carbon source (Findlay et al. 1986), meiofaunal species may be able to do so.
5. Meiofaunal species may be able to exert direct control over the pool size and renewal rate of some of their food resources.

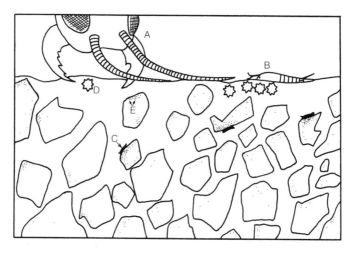

FIGURE 12.1. Schematic drawing of the sediment associated zoobenthos and its food resources. Consumers include a deposit feeding insect *(A)*, a copepod *(B)*, and some gastrotrichs *(C)*. Resources are algae *(D)*, and bacteria *(E)*, both of which are patchily distributed on a small scale, and of much higher nutritional quality than the sediment itself.

6. Omnivory, especially detritivory, may allow large benthic invertebrates to have higher population resistance stability (*sensu* Connell and Sousa 1983) than smaller, dietary specialists; when high quality particles are in low abundance, omnivores may persist on low quality (but high quantity) particles.

Few data are available to test any of these contentions, but some observations are consistent with the predictions. Some small benthic animals do appear to have extraordinarily specialized diets (Pourriot 1977), while many macroinvertebrates, e.g., gastropods, decapods, and insects, are omnivorous (cf. Pimm 1982).

In addition to affecting the relative feeding modes of large and small organisms, the physical structure of sediment may provide small, interstitial animals with a refuge from predators (Coull and Bell 1979). Although the few data now available suggest that most freshwater benthic animal production is consumed by predators (Strayer and Likens 1986), turnover rates of marine meiofaunal populations apparently are much lower than expected on the basis of body size (Banse and Mosher 1980), suggesting that these animals are subject to low rates of predation. Do lacustrine meiofauna have similarly low turnover rates? If so, does the physical structure of the sediment shelter them from predation, or do they possess other (behavioral, chemical, morphological) defenses against predators?

Landscape Ecology of the Littoral Zone

Patchiness on a scale readily perceived by humans is a conspicuous characteristic of the littoral zone. The description, causes, and consequences of patchiness in other habitats have received much attention from ecologists over the past decade (Paine and Levin 1981; Pickett and White 1985), and even more recently under the name landscape ecology (Forman and Godron 1986). Major themes in landscape ecology include the description of patchiness, the causes of patchiness, ecological interactions among patches, and the importance of the size, shape, and configuration of patches. Although most landscape ecologists are concerned with distances measured in kilometers, many of the patch attributes suggested by Forman and Godron (1981; Fig. 12.2) could usefully describe littoral zone habitat patches at smaller scales. The mosaic of habitats typical of littoral zones may be a good place to test the ideas put forth by landscape ecologists (e.g., by manipulating patch size, shape, or configuration, or by altering densities of key plant and animal species). Preliminary steps in this direction have been taken in the development of novel approaches to littoral zone management (Engel 1984, 1987).

Different patterns of habitat arrangement may accentuate or reduce interactions of organisms inhabiting different littoral zone substrates. For example, crayfish (*Orconectes* spp.) are restricted to rocky or vegetated substrates that provide refuge from fish predators (Stein and Magnuson

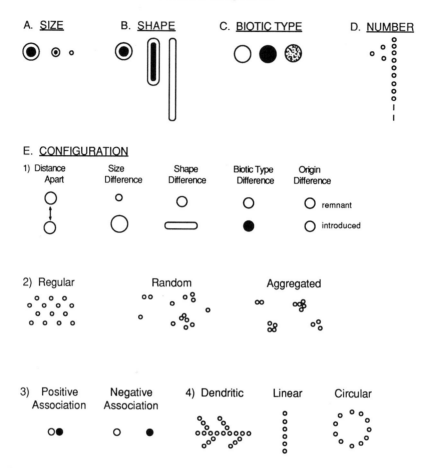

FIGURE 12.2. Attributes of landscape patches (from Forman and Godron 1981).

1976). During nocturnal feeding forays, though, crayfish may venture some distance from cover to feed on other invertebrates, algae, and macrophytes. Such activity patterns could produce grazing haloes similar to those resulting from grazing by urchins around coralheads (Ogden et al. 1973, Hay 1984). If, however, patches of macrophytes are interspersed with rocky patches, predictions are less certain. Crayfish may feed freely in macrophytes or, if fish predators of crayfish are associated with vegetation, the interaction may operate in the opposite direction. Fish ranging at the boundary of the vegetated habitat may produce a halo of behavioral inhibition of crayfish activity even within the rocky habitat. The effects of strong interactions may therefore be accentuated or inhibited, depending on the size and spatial arrangement of patches, and on direct and indirect interactions among specific food web components.

The Macrophyte-Periphyton-Grazer Complex

Most submersed surfaces are overgrown by periphyton—a structurally complex matrix of algae, bacteria, fungae, and associated detritus (Allanson 1973). Only a few of the many potentially important interactions surrounding the macrophyte- periphyton-grazer association have been explored (Table 12.1). Hutchinson (1975) and Thomas (1982) both speculated that there are symbiotic elements in the macrophyte-periphyton- grazer complex (reviewed in Carpenter and Lodge 1986), but these hypotheses have not been adequately tested for freshwater macrophytes.

We highlight five important questions surrounding the interactions among macrophytes, periphyton, and periphyton- grazers:

1. Under what conditions do macrophyte substrates contribute significantly to the nutrition of periphyton? We predict macrophyte products are proportionately more important to periphyton nutrition in oligotrophic systems where water-column nutrients are lower. The impact of macrophyte condition and temporal patterns of macrophyte senescence on macrophyte releases of dissolved organic and inorganic compounds should also be considered in any examination of macrophyte-periphyton interactions.
2. Do some macrophytes defend themselves chemically from periphyton?
3. Under what nutrient and light conditions does periphyton reduce the growth of macrophytes, and therefore under what natural conditions are periphyton grazers likely to be important in increasing macrophyte growth? We predict that periphyton would inhibit macrophyte growth, and grazers on periphyton would maximally enhance macrophyte growth, under low light conditions.
4. Whether a result of microclimate or specific products, to what extent do different macrophytes support different periphyton assemblages? How widespread is this phenomenon, and what are its implications for determining the distribution and abundance of selective grazers and their predators?
5. What determines the selectivity of grazers on periphyton—functional morphology of the grazer, morphological or chemical defenses of different algae, etc.? Are certain algae resistant to grazers? If so, what are the implications for their macrophyte substrates?

Grazing on Macrophytes

Grazing on living macrophytes has been known for years (Gaevskaya 1966), but its possible importance in community structure and ecosystem function has been underestimated in recent reviews (Gregory 1983; Wetzel 1983b). Important grazers of macrophytes in north temperate lakes include waterfowl (Anderson and Low 1976; Jupp and Spence 1977; Kiorboe 1981; Smith and Kadlec 1985), muskrats (Pelikan et al. 1971), fish (Prejs 1984),

TABLE 12.1. Beneficial (A) and detrimental (B) interactions among the components of the macrophyte–periphyton–grazer complex. Reference numbers are given in parentheses after each effect. Those effects without references have not, to our knowledge, been directly tested. Selected recent references, from both marine and freshwater systems, are listed below the table

A. Beneficial Effects (column benefits row)

	Macrophyte	Periphyton	Grazer
Macrophyte		Distract grazers	Remove periphyton[1] Recycle N,P[2] Provide CO_2
Periphyton	Illuminate substrate Release C,N,P[3]		Recycle N,P[4] Provide CO_2[4] Remove senescent peri.[4] Increase C,N,P release from macrophytes[5]
Grazer	Provide food[6] Provide predation refuge[7] Provide oviposition site Provide O_2 Remove NH_3 Provide attractants[8]	Provide food[9] Provide predation refuge Provide O_2 Remove NH_3	

B. Detrimental Effects (column harms row)

	Macrophyte	Periphyton	Grazer
Macrophyte		Shading[10] Reduce C,N,P diffusion[10] Increase drag	Consumption[6] Increase pathogens[5]
Periphyton	Antibiotic exudates[11]		Consumption[12] Increase sloughing[13]
Grazer	Chemical detractants	Chemical detractants	

[1]Rogers and Breen (1983); Orth and van Montfrans (1984); Brönmark (1985); Howard and Short (1986).

[2]Miura et al. (1978).

[3]McRoy and Goering (1974); Sondergaard (1983); Wetzel (1983a); Kirchman et al. (1984); Morin (1986); Burkholder (1986); cf. Cattaneo and Kalff (1978, 1979); Fontaine and Nigh (1983).

[4]Hunter and Russell-Hunter (1983); Lamberti and Moore (1984).

[5]Rogers and Breen (1983); Wallace and O'Hop (1985).

[6]Sheldon (1984); Carpenter and Lodge (1986); Lodge and Lorman (1987).

[7]Crowder and Cooper (1982); Gilinsky (1984); Leber (1985).

[8]Sterry et al. (1983).

[9]Lamberti and Moore (1984); Lodge (1986); Power (1987).

[10]Phillips et al. (1978); Cambridge et al. (1986); Sand-Jensen and Revsbech (1987); cf. Mazzela et al. (1986).

[11]Wium-Anderson et al. (1982).

[12]Cattaneo (1983); Lamberti and Moore (1984).

[13]Sumner and McIntire (1982).

and invertebrates (Sheldon 1984; Lodge and Lorman 1987). In tropical lakes, large mammals such as capybara and manatee are effective grazers on macrophytes. Reductions in macrophyte biomass by such common benthic organisms as crayfish (Lodge and Lorman 1987) and insects (Wallace and O'Hop 1985) may have large indirect impacts on macrophyte-associated organisms, which may ramify through the foodweb (Carpenter and Lodge 1986). The impact of these and other grazers, and the factors affecting macrophyte vulnerability, deserve much more work.

Four of the many exciting questions surrounding the interaction of macrophytes and macrophyte grazers are:

1. To what extent do macrophyte grazers reduce macrophyte standing crop, and what is their effect on macrophyte productivity?
2. What is the taxonomic distribution of macrophyte grazers, and how many are specialized feeders? We predict that many grazers (e.g., snails) (Sheldon 1984), are facultative grazers. When such grazers are at low density, they may eat only periphyton, but at high population densities, they may have a significant effect on macrophytes. Predators (e.g., fish) of these grazers may then indirectly control periphyton and macrophyte abundance.
3. Conversely, to what extent may grazers of macrophytes, by reducing macrophyte habitats, reduce the abundance of other species, including fish, that rely on macrophytes for feeding, refuge, or spawning?
4. What determines the selectivity of grazers? Are chemical defenses (Ostrofsky and Zettler 1986) widespread among macrophyte taxa? How important are structural and architectural defenses?

Predator-Prey Relations in the Littoral Zone

The importance of predation (especially by fish) in structuring littoral zone communities remains a vital question. It is clear that selective feeding by fish can profoundly affect invertebrate prey assemblages in both expected and surprising ways (reviewed by Sih et al. 1985; Thorp 1986).

Despite methodological difficulties, a number of experiments have produced substantial insight into the complex interactions between fish and littoral zone invertebrates. While fish predation often reduces total biomass and densities of larger prey (Crowder and Cooper 1982; Bohanan and Johnson 1983; Morin 1984a; Post and Cucin 1984; Brown and DeVries 1985), densities of other, smaller prey may be enhanced (Crowder and Cooper 1982; Gilinsky 1984; Morin 1984b). These enhanced species are themselves prey for the large, invertebrate predators (e.g., odonates) that are reduced by fish. Thus, for these smaller invertebrate species, any increased mortality resulting directly from fish predation is offset by the indirect release from predation by larger invertebrates. Such indirect effects, involving predation and competition, are probably more common and important than our present experimental designs have shown (Sih et al. 1985). They deserve much more attention from ecologists.

In addition, the roles of macrophytes and other habitat structures in modifying predator-prey interactions have only been outlined. For example, macrophytes provide a habitat and food source for fish, but also a hindrance to feeding activity at high density. The study by Crowder and Cooper (1982) suggests that bluegill (*Lepomis macrochirus*) feeding is most efficient at intermediate levels of vegetation density. A question of vital interest to both basic and applied ecologists is: What are the relationships among macrophyte density, fish predation and production, and invertebrate production?

The relationship between structure and fish foraging efficiency is not one-way. Through their nest building alone, many fishes reduce densities and change species composition of both macrophytes (Carpenter and McCreary 1985) and invertebrates (Cowell 1984). Under what conditions and for what species these nest-building disturbances are most important is unknown.

The extent to which parasites of fish and invertebrates affect the productivity, composition, and interactions among the benthos and their predators is also unknown (Minchella et al. 1985; Lodge et al. 1987; Price et al. 1986).

Aquatic Microclimates Created by Macrophytes

In addition to their importance in trophic interactions, macrophytes create very distinct aquatic microclimates that can affect the distribution of other organisms. Vertical and horizontal gradients in water temperature, pH, and dissolved oxygen in the littoral zone can be pronounced, relative to gradients in the epilimnetic zone (see review of Carpenter and Lodge 1986; Godshalk and Barko submitted). The pH and dissolved oxygen, examined over a midsummer diel period in dense beds of *Hydrilla* in the Potomac River, varied from 7-10 and from 3-17 mg/l, respectively (Barko et al., unpublished). Diel changes of these magnitudes must have far-reaching effects on the activity of animals, including fish, as well as on biogeochemical processes. Yet measurements of these diel physicochemical gradients have been rare, and their effects on trophic and other biological interactions have hardly been investigated.

Benthic-Pelagic Links

Biological and biogeochemical interactions within the littoral and profundal zones can have far-reaching effects on the pelagic zone. The biogeochemical interdependency of littoral and pelagic zones has been well documented by Wetzel (1979). Many other interhabitat links, however, especially those mediated by food web interactions, are less appreciated (Fig. 12.3). In this section, we suggest that a number of little studied interhabitat links may be critically important to the functioning of lake communities.

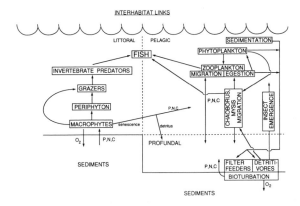

FIGURE 12.3. Simplified diagram of a lake community emphasizing the biological and biogeochemical links between habitats.

Nutrient Cycling

Investigations of the role of internal nutrient loading have significantly enhanced our understanding of lake eutrophication (Twinch and Peters 1984; Riley and Prepas 1984, 1985; Stauffer 1985; Ryding 1985). In the open water of lakes, nutrient cycling is influenced by many physical and chemical conditions, which affect microbial activity and particulate matter retention in the water column (Golterman 1975; Wetzel 1983). Seasonal changes in phytoplankton community composition and physiological condition of specific populations undoubtedly influence both the rate and direction of nutrient exchanges between open water and profundal sediments (Hecky and Kling 1981; Trimbee and Harris 1984). Seasonal variations in the ratios of available nutrients in the open water of lakes are driven in part by differential losses to profundal sediments. These processes, which may have important effects on phytoplankton composition, have only been partially investigated. In addition, the vast majority of nutrient cycling studies have concentrated on profundal to open water processes, with relatively little attention directed towards the littoral zone, despite evidence that this region can be a major source of nutrient input to the epilimnetic zone (Prentki et al. 1979; Carpenter 1980; Smith and Adams 1986).

Nutrient uptake from sediments by rooted aquatic macrophytes constitutes a potential net transfer of nutrients towards the water. The magnitude of this transfer can be prodigious, and is nearly a linear function of macrophyte growth rate (Barko and Smart 1986). Nutrients are released from aquatic macrophytes, primarily during senescence and decay (Carpenter 1980; Denny 1980; Barko and Smart 1981). These nutrients may then reenter the sediment as detritus or in the form of inorganic precipitates, may be incorporated locally by microorganisms, or may be transported to the open water. The relative partitioning of nutrients into these potential sinks and how the partitioning varies among lakes is unknown.

These effects need to be evaluated in whole system investigations, with greater attention to lakewide nutrient source-sink relationships.

Changes in Nutrient Ratios

A potentially very important aspect of nutrient mobilization from littoral sediments by rooted aquatic macrophytes is the possibility that this transfer might cause pelagic nutrient imbalances. Given the stoichiometry of nutrient release from aquatic macrophytes during decay (with enriched release of P relative to N relative to C), in combination with progressive concentration of N in macrophyte detritus through time (Godshalk and Wetzel 1978; Twilley et al. 1986), it is probable that proportionately more soluble P than soluble N is exported from the littoral zone. This trend may be intensified by nitrogen loss and transformation at the sediment surface (Reddy and Patrick 1984). However, sediment oxidation by macrophytes may enhance P retention by sediments (Jaynes and Carpenter 1986). The net result of these interacting processes has not been tested. We predict that there is a reduction in the ratio of N to P in water transported from the littoral to the pelagic region of lakes, which may stimulate the development of N-fixing blue-green algae. Investigations of hydraulic exchange and associated nutrient fluxes between littoral and pelagic regions of lakes should be particularly instructive.

Predator-Prey Relations

Many of the fish species (e.g., *Lepomis* spp., *Perca flavescens*) featured in littoral zone predator-prey experiments would ordinarily provide a link between littoral and pelagic habitats. Adults of these and piscivorous species commonly move between the littoral and pelagic zones, often feeding in one area and resting in the other. In marine systems, such fish movements transport significant quantities of nutrients between habitats (Meyer and Schultz 1985). Perhaps more importantly, many fishes undergo ontogenic niche shifts from open water planktivory to benthivory or piscivory in the littoral zone (Werner and Gilliam 1984). These periods of planktivory have major impacts on pelagic community structure and productivity (Mills and Forney this volume; Carpenter et al. 1987). Conversely, the abundance and quality of planktonic food for young of the year and juvenile fish partly determines the success of recruitment to the littoral adult stage.

The density of piscivorous fish in lakes also influences the extent of fish migrations between littoral and pelagic habitats. Small fish may remain in the safety of vegetated habitats when piscivores are present (Werner et al. 1983, Mittelbach 1986). Thus, the link between habitats resulting from fish migrations is a dynamic one, often controlled by the relative risks and rewards associated with each habitat. Addition or removal of piscivores from a system (i.e., biomanipulation *sensu* Shapiro and Wright 1984) can therefore affect not only the density of prey fishes in lakes but also their movements between habitats and thus the link between littoral and pelagic zones.

Links *via* fishes between the profundal and pelagic zones are also apparent, but their importance is even less well understood than the littoral-pelagic links. Many fishes that feed predominantly in open water also forage on profundal benthos (including fishes)—either on the bottom or as benthic organisms migrate up into the water column. The importance of this alternative food source for pelagic fishes, and the impact of fish predation on profundal community structure require much further study.

Vertical Migrations

Several prominent species (*Chaoborus*, *Mysis*, many cyclopoid copepods, *Melosira*, and some blue-green algae) spend part of their lives on the sediments and part in the pelagic zone. These species can carry materials or information from one habitat to another and more importantly, the distribution and abundance of meroplanktonic species can be affected by events that occur in either habitat. For example, although *Chaoborus* usually is thought of as a planktivore, Jonasson (1972) has suggested that benthic feeding may support *Chaoborus* through the winter in Lake Esrom, when the density of zooplankton prey is low. Similarly, blooms of some blue-green algae may begin on the sediments, where the population takes up key elements such as iron before moving up into the open water.

Zooplankton and fish can also have major effects on nutrient movements across habitat boundaries. Interspecific differences in the mechanism of grazing by zooplankton, and associated effects on phytoplankton, may differentially affect nutrient cycling and nutrient ratios in the pelagic and profundal zones. Unconsolidated zooplankton excrement can be mineralized, with nutrients returned directly to the water column. In contrast, dense fecal pellets formed by some zooplankton (e.g., copepods) sink rapidly and may thus fertilize the profundal benthos.

In north temperate lakes, crayfish, snails, and probably other invertebrates migrate into deeper water in winter. Conversely, some insect larvae migrate from deep to shallow waters prior to emergence. Emergence of benthic insects carries nutrients from the sediments to the surface waters and surrounding landscape. Insect emergence can represent a substantial loss of key nutrients from a lake (Walter 1985). Also, insect emergence provides a pathway by which nutrients can be returned from the sediments to the epilimnion. For example, an emergence of 1 g $DW \cdot m^{-2}$ carries approximately 5-10 mg $P \cdot m^{-2}$ into the epilimnion. To date, the community and ecosystem consequences of these cross-habitat interactions have not been fully explored.

Transport of Detritus

Depending on the origin of littoral zone detritus, the extent of its decomposition, and physical factors affecting transport, the quality and quantity of littoral detritus entering the pelagic and profundal zones varies among lakes. As mineralization in sediments proceeds, reduced products accumulate in soluble forms with effects not only on animal use, but on nutrient cycling and lake metabolism (Rich and Wetzel 1978; Wetzel 1979). The quality and quantity of detritus reaching the profundal benthos depends not only on littoral zone sources, but also on the productivity and species composition of the phytoplankton, and on the respiratory demands and species composition of the planktonic consumers (microbes vs fecal pellet forming zooplankton vs nonfecal pellet forming zooplankton), as well as the depth and mixing regime of the pelagic zone.

Bioturbation

The feeding and burrowing activities of benthic invertebrates (especially the tubificid oligochaetes) can increase the release of nitrogen and phosphorus from sediments into the overlying water (Fukuhara and Sakamoto 1987). Although the effects of bioturbation on sediment properties have been explored in some detail (Robbins 1982; McCall and Fisher 1980), we do not know to what extent events in the pelagic zone are affected. It seems reasonable to expect this link to be most significant in shallow, well mixed lakes, where sediments are in direct contact with the euphotic zone.

Transport of Algae

For unknown reasons, phytoplankton species composition and productivity in the littoral zone appears to be very different from that in the open water of lakes (Kairesalo 1980, Barko et al. 1984). The relative extent to which dislodgement of algae from macrophytes, grazing pressures, nutrient supply, and competitive processes are involved in promoting these differences is ripe for investigation.

   In addition, it is possible that chemical and biological interactions between the littoral, profundal, and pelagic zones may influence shifts in algal species assemblages (Lemly and Dimmick 1982; Kairesalo and Koskimies 1985). For example, the importance of the profundal zone in annually providing inocula of many algal groups to the pelagic zone has not been tested (Reynolds 1984, pp. 217-221).

Benthic Filter Feeders

Benthic filter feeders may consume seston and compete with zooplankton. In many lakes, the dominant benthic filter feeders filter a relatively small volume of water (i.e., less than the epilimnetic volume per year; Strayer et al. 1981; James 1987). However, in other lakes, benthic filter feeders such as sponges or the zebra mussel *Dreissena polymorpha* may develop massive populations and consume immense quantities of seston (Stanczykowska et al. 1976; Frost 1978; Lewandowski 1983).

Factors Affecting the Strength of Interhabitat Links

Many of the controversies among community ecologists (e.g., the importance of fish predation on littoral invertebrates) result from a desire for either/or answers. Questions may be couched more profitably in terms of relative importance and gradients affecting importance. With that in mind, we will outline below a series of speculations about how the relative importance of the interactions previously described changes along gradients of lake morphometry, seasonality, trophic status, and community composition.

Lake Morphometry

The importance of biogeochemical couplings among the littoral, profundal, and pelagic zones can be at least partially predicted on the basis of lake morphometry (Carpenter 1983). How biological links differ in relation to lake morphometry is, however, less well understood.

Many of the biotic interactions between the littoral and pelagic habitats are probably a function of distance between the two habitats. Yellow perch, which attach their eggs to littoral zone substrates, are abundant in both Oneida Lake and Lake Michigan. In Oneida Lake (see Mills and Forney this volume) perch young of the year have a dramatic influence on pelagic zooplankton. Yet the same fish species in Lake Michigan has little, if any, effect in the central part of Lake Michigan (Scavia and Fahnenstiel this volume). We might therefore expect the habitat link *via* fishes to be greatest in smaller lakes and lakes with finely dissected shorelines (e.g., southern reservoirs) where distances between habitats are less.

The physical structure of littoral zone habitats may also affect the fish link between littoral and pelagic zones. While a littoral zone devoid of structure may not support populations of some fish species, even a small patch of macrophytes or other littoral zone structure may provide the necessary spawning habitat to inoculate the pelagic zone with large numbers of planktivorous young of the year fishes. Given the individual fecundity of most fishes, a few adult fish can produce a large number of larvae. Testing the relationship between the macrophyte landscape and the community structure of fishes should be a priority for ecologists and of great interest to lake managers.

Seasonality of Macrophyte Abundance

Food Web Effects

Temporal patterns of the occurrence of macrophytes in the littoral zone may have a major effect on littoral-pelagic links. The seasonality of macrophyte growth and its relationship to the timing of invertebrate and fish reproduction, in particular, may affect the strength of food web interactions within the littoral and the significance of the fish inoculum from littoral to pelagic zone. We might expect that in lakes with little seasonality (i.e., with more nearly permanent littoral structure and higher year-round temperatures and metabolic rates), the importance of the fish link between littoral and pelagic zones would increase.

While many tropical lakes have strong wet-dry seasonality that affects macrophyte growth, other tropical and subtropical lakes have a year-round growing season for macrophytes, a pattern that does not typically occur in temperate lakes. As a consequence of the near constant food source, we might also expect a greater diversity of herbivores (especially obligate herbivores) in tropical and other nonseasonal lakes. This expectation is supported for nearshore marine fishes (Gaines and Lubchenco 1982) and partly borne out for freshwater lentic fishes. For freshwater fishes, there are few macrophyte consumers in North America, but many in the low latitudes of South America, Africa, and Asia. This temperate-tropical comparison, however, is not without obvious exceptions. There are many, common species of freshwater herbivorous fishes in temperate Europe (Prejs, 1984). The relationships between temporal patterns of resource abundance, food web structure, and biogeographical constraints on community structure clearly require further documentation and testing.

Effects on Nutrient Cycling

Predictions about trends in the importance of macrophytes as nutrient regenerators are more difficult to make. With increasing constancy of macrophyte abundance and greater annual macrophyte production in less seasonal, warm environments (cf. Barko and Smart 1981), the importance of macrophytes as nutrient pumps may become more important. Senescence and concomitant release of nutrients would be more constant than in seasonal environments and probably greater on an annual time scale. The net effect of greater macrophyte production and polymixis on the nutrient link between littoral and pelagic zones is difficult to predict.

Macrophytes that senesce and thereby release nutrients continuously or in a pulsed fashion during the summer probably have a greater impact on phytoplankton production than those that senesce only during the fall (Carpenter 1980; Smith and Adams 1986). Senescence in some submersed macrophyte species (exotics in particular) can be nearly continuous, particularly in eutrophic systems (Barko and Smart 1980). In reality, however, very little information is available to allow an assessment of the timing and associated impacts of different macrophyte species on sediment to overlying water nutrient cycling.

Lake Trophic Status

On the continuum from oligotrophy to eutrophy, macrophyte biomass is probably unimodal, reaching a peak somewhere near the middle of the scale. The relative importance of the littoral zone as a nutrient source probably follows the same trajectory (Wetzel 1979, Carpenter 1981). At very high productivity, shading by phytoplankton or periphyton reduces macrophyte biomass, and the importance of nutrient regeneration from the littoral zone may decline. We also predict that the importance of the fish link between littoral and pelagic zones follows the macrophyte biomass curve, but this idea has not been tested.

Species Composition

The widespread importance of strong interactions (*sensu* Paine 1980) and indirect effects (Sih et al. 1985; Kerfoot and Sih 1987) in aquatic communities suggests that interhabitat links may be very sensitive to species composition. While replacing some species may have little effect, changing others may have a tremendous effect. Learning how to predict these effects should be a priority for lacustrine ecologists and lake managers.

Macrophyte Additions and Deletions

The different microclimates created by different vegetation types—floating, floating-leaved, submersed, and emergent—may foster very different littoral communities and have very different impacts on both the nutrient and fish links between habitats. Floating, floating-leaved, and emergent macrophytes, much more than most submersed species, all substantially decrease light penetration, water column temperature, photosynthesis, and dissolved oxygen. Oxygenation of sediments by rooted macrophytes, can substantially elevate sediment redox potential (Tessenow and Baynes 1978; Carpenter et al. 1983; Jaynes and Carpenter 1986), but variations in this ability among submersed species is great (Sand-Jensen et al. 1982). With rooted macrophytes, these effects are localized to shallow areas, but floating macrophytes can produce lakewide effects.

The dramatic increase of macrophyte biomass that often accompanies invasions of exotic species probably also has far-reaching effects on nutrient recycling and biotic interactions. We predict that increases in macrophyte biomass would lead to increased littoral and pelagic productivity through both enhanced phosphorus recycling from sediments and through the food web effects of enhanced recruitment of fishes (Fig. 12.4, Carpenter and Lodge 1986). Enhanced fish recruitment would increase grazing on zooplankton and decrease grazing on phytoplankton. Harvesting macrophytes would probably reverse these postulated effects.

Consumer Additions and Deletions

Experimental or accidental introduction of fishes has shown the dramatic effect of adding one species to lake systems (Zaret and Paine 1973). Decreases in pelagic primary production following introduction of grass carp (Mitzner 1978) suggest that carp reduce the importance of the nutrient link between littoral and pelagic zones (Mitzner 1978, reviewed in Carpenter and Lodge 1986).

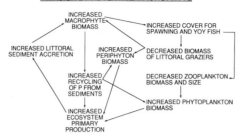

FIGURE 12.4. Hypothesized responses of a lake community to increased macrophyte biomass during invasion by an adventive species. YOY stands for young of the year.

The impact of grass carp on the fish link and planktonic community remains unclear. Any reduction in fish with zooplanktivorous larvae, though, would simply augment the effects of reduced nutrient cycling *via* macrophytes. The potential for similar effects with an introduced crayfish is clear (Fig. 12.5; Carpenter and Lodge 1986; Lodge and Lorman 1987). Yet different outcomes are not hard to imagine, depending on the configuration of the original food web.

Imagine a food web consisting initially of a piscivorous fish (with zooplanktivorous larvae), a zooplanktivorous cyprinid, plus the littoral food web. A reduction in macrophyte biomass by grass carp or crayfish may reduce nutrient regeneration by macrophytes and therefore reduce planktonic production. The piscivore population may decline as a result of the loss of habitat. Zooplanktivory by larval piscivores will then decline, but might be more than compensated for by increases in the population of the zooplanktivorous fish. Consequent reductions in zooplankton might enhance phytoplankton production. Whether the positive or negative effect of macrophyte removal on pelagic production would be more important would be very difficult to predict.

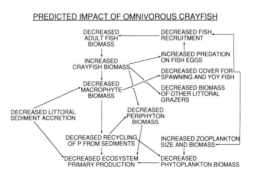

FIGURE 12.5. Hypothesized responses of a lake community to an irruption of omnivorous crayfish. YOY stands for young of the year.

Research Needs and New Experimental Approaches

In this chapter, we have highlighted many untested questions about benthic ecology and interhabitat interactions. Some of these questions (e.g., those involving the microbial component, the meiofaunal component, and littoral landscapes) are largely speculative, but they may turn out to be vital to understanding the function of aquatic communities. Other questions (e.g., involving the effect of decomposition on nutrient ratios and phytoplankton species composition, the macrophyte-periphyton-grazer complex, and the impact of macrophyte grazers on ecosystem function) have their basis in previously completed, small-scale research. The latter topics need much more work, especially on a spatial scale adequate for rigorously testing the comprehensive question of the function of whole-lake communities. Below, we outline some scientific and institutional limitations to achieving this synthetic understanding of lake communities, and propose some possible solutions.

Need for New Research Tools

Sampling and Processing Methods

There are good reasons for conducting research in the pelagic zone, a chief one being that open-water research often is logistically simpler than work on sediments or vegetation. Current methods for collecting and processing benthic samples are labor-intensive, and often imprecise and primitive (Downing and Anderson 1985, Downing and Cyr 1985). Improvement and development of methodology should be a priority over the next decade. Finally, both funding agencies and open-water limnologists need to recognize that a large amount of labor is often needed to conduct worthwhile research in the littoral and profundal zones. Labor expenses are as much a part of this research as, for example, a fluorescence microscope or image analyzer is to a microbial ecologist.

Remote Sensing

When Landsat I was launched in 1972, ecologists were offered a new view of the Earth. Initial uses of the imagery included lake recognition (Boland et al. 1979) and measurement of trophic status (Boland 1976; Almanza and Melack 1985) and turbidity (Holyer 1978; Strong and Eadie 1978). Subsequently, there have been considerable improvements in the spectral and spatial resolution of satellite sensors now in orbit or to be flown in the next decade (Butler 1984).

It is now possible with Landsat's Thematic Mapper (TM) to examine lakes with a spatial resolution of 30 m and with sufficient spectral information to recognize a range in concentrations of phytoplankton or suspended particulates and to decipher fringing plant communities. Digital analysis of the imagery permits correction for atmospheric interference and enhancement of features of interest. In regions where clouds are not frequent, repetitive coverage every 16 days is possible. One TM image can provide complete coverage of many lakes. Hence, one can determine both the within-lake extent of littoral vegetation or patchiness in phytoplankton abundance and the between-lake variance of these characteristics. In regions where neighboring lakes are similar in most physical and chemical aspects, the effects of food web differences on algal or macrophyte abundance could be tested. The regional sampling provided by satellite imagery can then guide further ground-based study.

Large-Scale Experiments

Many interactions highlighted in this chapter have not been tested. Some are amenable to typical laboratory and field experiments. For many, however, the crucial experiments have not been performed because an adequate experimental design would require large multidisciplinary teams, unusually large experimental sites, and long periods of time. While pelagic mesocosms of manageable size are many times the size of zooplankton and phytoplankton patches, cages for littoral zone experiments often are not large enough to enclose a sufficient number of organisms or substrate patches. Certainly, most questions involving habitat interactions cannot be tested using experiments of conventional spatial scale. To test habitat links in an acceptably realistic way may require multiyear experiments in ponds or whole lakes. Experiments on that spatial and temporal scale and including botanical and zoological ecologists, physical limnologists, hydrologists, etc. are not often funded by the Ecology Program of the National Science Foundation (NSF).

However, large-scale manipulations are initiated every year all over the U.S. by lake management associations and state departments of natural resources. Piscivore manipulations and grass carp introductions are commonplace, and provide tremendous opportunities for community ecologists to study many of the habitat links described in this chapter. Unfortunately, there is often little communication between government agencies, resource managers, and academic ecologists (Barko et al. 1986). Better local efforts may be a solution, but a national clearinghouse of information on upcoming manipulations, perhaps supported by NSF, might be helpful. Plans for any large-scale experiment need to be made well in advance so that careful consideration can precede collection of at least one year of premanipulation data. On the other hand, NSF needs to be able to respond quickly so that ecologists may take advantage of manipulations planned by resource managers.

Most of the world's lakes are small and vegetated, the sort of lakes where interhabitat interactions are likely to be most important. Many of these lakes are heavily used by humans for both sustenance and recreation. Questions about the function of littoral, profundal, and pelagic habitats and their interactions are much too important for both basic and applied ecology to remain untested.

References

Allanson, B. R. 1973. The fine structure of the periphyton of *Chara* sp. and *Potamogeton natans* from Wytham Pond, Oxford, and its significance to the macrophyte-periphyton metabolic model of R. G. Wetzel and H. L. Allen. Freshwat. Biol. 3:535–541.

Almanza, E. and J. M. Melack. 1985. Chlorophyll differences in Mono Lake (California) observable on Landsat imagery. Hydrobiologia 122:13–17.

Anderson, M. G. and J. P. Low. 1976. Use of sago pondweed by waterfowl on the Delta Marsh, Manitoba. J. Wildl. Manage. 40:233–242.

Banse, K. and S. Mosher. 1980. Adult body mass and annual production/biomass relationships of field populations. Ecol. Monogr. 50:355–379.

Barko, J. W., M. S. Adams and N. L. Clesceri. 1986. Environmental factors and their consideration in the management of submersed vegetation: A review. J. Aquat. Plant Manage. 24:1–10.

Barko, J. W., D. J. Bates, G. J. Filbin, S. M. Hennington, and D. G. McFarland. 1984. Seasonal growth and community composition of phytoplankton in a eutrophic Wisconsin impoundment. J. Freshwat. Ecol. 2:519–533.

Barko, J. W. and R. M. Smart. 1980. Mobilization of sediment phosphorus by submersed freshwater macrophytes. Freshwat. Biol. 10:229–238.

Barko, J. W. and R. M. Smart. 1981a. Comparative influences of light and temperature on growth and metabolism of selected submersed freshwater macrophytes. Ecol. Monogr. 51:219–235.

Barko, J. W. and R. M. Smart. 1981b. Sediment-based nutrition of submersed macrophytes. Aquat. Bot. 10:339–352.

Barko, J. W. and R. M. Smart. 1986. Sediment-related mechanisms of growth limitation in submersed macrophytes. Ecology 67:1328–1340.

Bohanan, R. E. and D. M. Johnson. 1983. Response of littoral invertebrate populations to a fish exclusion experiment. J. Freshwat. Invert. Biol. 2:28–40.

Boland, D. H. P. 1976. Trophic classification using Landsat I multispectral scanner data. USEPA. 600/3-76-037.

Boland, D. H. P. et al. 1979. Trophic classification of selected Illinois water bodies. EPA. 600/3-79-123.

Brönmark, C. 1985. Interactions between macrophytes, epiphytes and herbivores: An experimental approach. Oikos 45:26–30.

Brown, K. M. and D. Devries. 1985. Predation and the distribution and abundance of a pond snail. Oecologia 66:93–99.

Burkholder, J. M. 1986. Seasonal dynamics, alkaline phosphatase activity and phosphate uptake of adenate and loosely attached epiphytes in an oligotrophic lake. Ph.D. dissertation. Michigan State University, East Lansing. 317 pp.

Butler, D. M., Chairman. 1984. Earth observing system. Science and mission requirements working group report. NASA Tech. Memorandum 86129. Goddard Space Flight Center, Green Belt, Maryland.

Cambridge, M. L., A. W. Chiffings, C. Brittan, L. Moore and A. J. McComb. 1986. The loss of seagrass in Cockburn Sound, Western Australia. II. Possible causes of seagrass decline. Aquat. Bot. 24:269–285.

Carpenter, S. R. 1980. Enrichment of Lake Wingra, Wisconsin, by submersed macrophyte decay. Ecology 61:1145–1155.

Carpenter, S. R. 1981. Submersed vegetation: An internal factor in lake ecosystem succession. Am. Nat. 118:372–383.

Carpenter, S. R. 1983. Lake geometry: Implications for production and sediment accretion rates. J. Theor. Biol. 105:273–286.

Carpenter, S. R., J. J. Elser, and K. M. Olson. 1983. Effects of roots of *Myriophyllum verticillatum L.* on sediment redox conditions. Aquat. Bot. 17:243–249.

Carpenter, S. R., J. F. Kitchell, J. R. Hodgson, P. A. Cochran, J. J. Elser, M. M. Elser, D. M. Lodge, D. Kretchmer, X. He, and C. N. von Ende. 1987. Regulation of lake ecosystem primary productivity by food web structure in whole-lake experiments. Ecology 68:1863–1876.

Carpenter, S. R. and D. M. Lodge. 1986. Effects of submersed macrophytes on ecosystem processes. Aquat. Bot. 26:341–370.

Carpenter, S. R. and N. J. McCreary. 1985. Effects of fish nests on pattern and diversity of submersed vegetation in a softwater lake. Aquat. Bot. 22:21–32.

Cattaneo, A. 1983. Grazing on epiphytes. Limnol. Oceanogr. 28:124–132.

Cattaneo, A. and J. Kalff. 1978. Seasonal changes in the epiphyte community of natural and artificial macrophytes in Lake Memphremagog (Que.-Vt.). Hydrobiologia 60:135–166.

Cattaneo, A. and J. Kalff. 1979. Primary production of algae growing on natural and artificial aquatic plants: A study of interactions between epiphytes and their substrate. Limnol. Oceanogr. 24:1031–1037.

Connell, J. H. and W. P. Sousa. 1983. On the evidence needed to judge ecological stability or persistence. Am. Nat. 121: 789–824.

Coull, B. C. and S. S. Bell. 1979. Perspectives of marine meiofaunal ecology. *In*: Ecological processes in coastal and marine systems, ed. R. J. Livingston, 189–216. New York: Plenum.

Cowell, B. C. 1984. Benthic invertebrate recolonization of small-scale disturbances in the littoral zone of a subtropical Florida lake. Hydrobiologia 109:193–205.

Crowder, L. B. and W. E. Cooper. 1982. Habitat structural complexity and the interaction between bluegills and their prey. Ecology 63:1802–1813.

Denny, P. 1980. Solute movement in submerged angiosperms. Biol. Rev. 55:65–92.

Downing, J. A. and M. R. Anderson. 1985. Estimating the standing biomass of aquatic macrophytes. Can. J. Fish. Aq. Sci. 42:1860–1869.

Downing, J. A. and H. Cyr. 1985. Quantitative estimation of epiphytic invertebrate populations. Can. J. Fish. Aq. Sci. 42:1570–1579.

Engel, S. 1984. Restructuring littoral zones: A different approach to an old problem. *In:* Lake and reservoir management, 463–466. Proc. 3rd Ann. Conf. North Am. Lake Manage. Soc., Oct. 18-20, 1983, Knoxville, Tenn. EPA Rep. No. 440/5-84-001. U.S. Environ. Prot. Agency, Washington, D.C.

Engel, S. 1987. The restructuring of littoral zones. Lake Reservoir Manage. 3:235–242. Proc. North Am. Lake Manage. Soc., Washington, D.C.

Fee, E. J. 1979. A relation between lake morphometry and primary productivity and its use in interpreting whole-lake eutrophication experiments. Limnol. Oceanogr. 24:401–416.

Findlay, S., J. L. Meyer, and P. J. Smith. 1986. Incorporation of microbial biomass by *Peltoperla* sp. (Plecoptera) and *Tipula* (Diptera). J. N. Am. Benthol. Soc. 5:306–310.

Folt, C. 1987. An experimental analysis of costs and benefits of zooplankton aggregation. In: Predation. Direct and indirect impacts on aquatic communities, ed. W. C. Kerfoot and A. Sih, 300–314. Hanover: University Press of New England.

Fontaine, T. D. III and D. G. Nigh. 1983. Characteristics of epiphyte communities on natural and artificial submersed lotic plants: Substrate effects. Arch. Hydrobiol. 96(3):293–301.

Forman, R. T. T. and M. Godron. 1981. Patches and structural components for a landscape ecology. BioScience 31:733–740.

Forman, R. T. T. and M. Godron. 1986. Landscape ecology. New York: John Wiley & Sons.

Frost, T. M. 1978. Impact of the freshwater sponge Spongilla lacustris on a Sphagnum bog-pond. Verhandlungen der internationalen Vereinigung fur Limnologie 20:2368–2371.

Fukuhara, H. and M. Sakamoto. 1987. Enhancement of inorganic nitrogen and phosphate release from lake sediment by tubificid worms and chironomid larvae. Oikos 48:312–320.

Gaevskaya, N. W. 1966. The role of higher aquatic plants in the nutrition of animals in freshwater basins. Moscow. (N.L.L. translation by D. G. Maitland; K. H. Mann, ed.), 3 vols.

Gilinsky, E. 1984. The role of fish predation and spatial heterogeneity in determining benthic community structure. Ecology 65:455–468.

Godshalk, G. L. and R. G. Wetzel. 1978. Decomposition in the littoral zone of lakes. In: Freshwater wetlands: Ecological processes and management potential, ed. R. E. Good, D. F. Whigham, and R. L. Simpson, 131–143. New York: Academic Press.

Gregory, S. V. 1983. Plant-herbivore interactions in stream systems. In: Stream ecology, ed. J. R. Barnes and G. W. Minshall, 157–189. New York: Plenum.

Haffner, G. D., D. J. Poulton, and B. Kohli. 1982. Physical processes and eutrophication. Water Res. Bull. 18:457–464.

Hay, M. E. 1984. Patterns of fish and urchin grazing on Caribbean coral reefs: Are previous results typical? Ecology 65:446–454.

Heaney, S. I., W. J. P. Smyly, and J. F. Talling. 1986. Interactions of physical, chemical and biological processes in depth and time within a productive English lake during summer stratification. Int. Revue Ges. Hydrobiol. 71:441–494.

Hecky, R. E. and H. J. Kling. 1981. The phytoplankton and protozooplankton of the euphotic zone of Lake Tanganyika: Species composition, biomass, chlorophyll content, and spatio-temporal distribution. Limnol. Oceanogr. 26:548–564.

Holyer, R. J. 1978. Toward universal multispectral suspended sediment algorithms. Remote Sens. Environ. 7:323–338.

Howard, R. K. and F. T. Short. 1986. Seagrass growth and survivorship under the influence of epiphyte grazers. Aq. Bot. 24:287–302.

Hunter, R. D. and W. D. Russell-Hunter. 1983. Bioenergetic and community changes in intertidal Aufwuchs grazed by Littorina littorea. Ecology 64:761–769.

Hutchinson, G. E. 1975. A treatise on limnology, vol. III. Limnological botany. New York: Wiley, 660 pp.

James, M. R. 1987. Ecology of the freshwater mussel Hydridella menziesi (Gray) in a small oligotrophic lake. Archiv fur Hydrobiologie 108:337–348.

Jaynes, M. L. and S. R. Carpenter. 1986. Effects of vascular and nonvascular macrophytes on sediment redox and solute dynamics. Ecology 67:875–882.

Jonasson, P. M. 1972. Ecology and production of the profundal benthos in relation to phytoplankton in Lake Esrom. Oikos Suppl. 14:1–148.

Jupp, B. P. and D. H. N. Spence. 1977. Limitations of macrophytes in a eutrophic lake, Loch Leven, II. Wave action, sediments and waterfowl grazing. J. Ecol. 65:431–446.

Kairesalo, T. 1980. Diurnal fluctuations within a littoral plankton community in oligotrophic Lake Paajarvl, Southern Finland. Freshwat. Biol. 10:533–547.

Kairesalo, T. and I. Koskimies. 1985. Vernal succession of littoral and nearshore phytoplankton: Significance of interchange between the two communities. Aqua Fennica 15:115–126.

Kerfoot, C. W. and A. Sih, eds. 1987. Predation: Direct and indirect impacts on aquatic communities. Hanover: University Press of New England. 386 pp.

Kiorboe, T. 1980. Distribution and production of submerged macrophytes in Tipper Grund (Ringk bing Fjord, Denmark), and the impact of waterfowl grazing. J. Appl. Ecol. 17:675–688.

Kirchman, D. L., L. Mazzella, R. S. Alberte and R. Mitchell. 1984. Bacterial epiphytes on *Zostera marina*. Mar. Ecol. Prog. Ser. 15:207–211.

Lamberti, G. A. and J. W. Moore. 1984. Aquatic insects as primary consumers. *In*: The ecology of aquatic insects, ed. V. H. Resh and D. M. Rosenberg, 164–195. New York: Praeger.

Landers, D. H. 1982. Effects of naturally senescing aquatic macrophytes on nutrient chemistry and chlorophyll a of surrounding waters. Limnol. Oceanogr. 27:428–439.

Leber, K. M. 1985. The influence of predatory decapods, refuge and microhabitat selection on seagrass communities. Ecology 66:1951–1964.

Lemly, A. D. and J. F. Dimmick. 1982. Phytoplankton communities in the littoral zone of lakes: Observations on structure and dynamics in oligotrophic and eutrophic systems. Oecologia 54:359–369.

Lewandowski, K. 1983. Occurence and filtration capacity of young plant-dwelling *Dreissena polymorpha* (Pall.) in Majcz Wielki Lake. Polskie Archiwum Hydrobiologii 30:255–262.

Lodge, D. M. 1986. Selective grazing on periphyton: A determinant of freshwater gastropod microdistributions. Freshwat. Biol. 16:831–841.

Lodge, D. M., K. M. Brown, S. P. Klosiewski, R. A. Stein, A. P. Covich, B. K. Leathers and C. Brönmark. 1987. Distribution of freshwater snails: Spatial scale and the relative importance of physiochemical and biotic factors. Am. Malacological Bull. 5:73–84.

Lodge, D. M. and J. G. Lorman. 1987. Reductions in submersed macrophyte biomass and species richness by the crayfish *Orconectes rusticus*. Can. J. Fish. Aq. Sci. 44:591–597.

Lopez, G. R. and I. J. Holopainen. 1987. Interstitial suspension-feeding by *Pisidium* spp. (Pisidiidae: Bivalvia): a new guild in the lentic benthos? Amer. Malacol. Bull. 5:21–30.

Mazzela, L. and R. S. Alberte. 1986. Light adaptation and the role of autotrophic epiphytes and primary production of the temperate seagrass, *Zostera marina* L. J. Exp. Mar. Biol. Ecol. 100:165–180.

McCall, P. L. and J. B. Fisher. 1980. Effects of tubificid oligochaetes on physical and chemical properties of Lake Erie sediments. *In*: Aquatic oligochaete biology, ed. R. O. Brinkhurst and D. G. Cook, 253–317. New York: Plenum.

McIntyre, A.D. 1969. Ecology of marine meiobenthos. Biol. Rev. 44:245–290.

McRoy, C. P. and J. J. Goering. 1974. Nutrient transfer between seagrass *Zostera marina* and its epiphytes. Nature 248:173–174.

Meyer, J. L. and E. T. Schultz. 1985. Migrating haemulid fish as a source of nutrients and organic matter on coral reefs. Limnol. Oceanogr. 30:146–156.

Minchella, D. J., B. K. Leathers, K. M. Brown, and J. N. McNair. 1985. Host and parasite counter adaptations: An example from a freshwater snail. Am. Nat. 126:843–854.

Mittelbach, G. 1986. Predator-mediated habitat use: Some consequences for species interactions. Env. Biol. Fish. 16:159–169.

Mitzner, L. 1978. Evaluation of biological control of nuisance aquatic vegetation by grass carp. Trans. Am. Fish Soc. 107:135–145.

Miura, T., K. Tanimizu, Y. Iwosa, and A. Kawakita. 1978. Macroinvertebrates as an important supplier of nitrogenous nutrients in a dense macrophyte zone in Lake Biwa. Verh. Int. Verein Limnol. 20:1116–1121.

Morin, J. O. 1986. Initial colonization of periphyton on natural and artificial apices of *Myriophyllum heterophyllum* M. zhx. Freshwat. Biol. 16:685–694.

Morin, P. J. 1984a. The impact of fish exclusion on the abundance and species composition of larval odonates: Results of short-term experiments in a North Carolina farm pond. Ecology 65:53–60.

Morin, P. J. 1984b. Odonate guild composition: Experiments with colonization history and fish predation. Ecology 65:1866–1873.

Ogden, J. C., R. A. Brown, and N. Salesky. 1973. Grazing by the echinoid *Diadema antillarum*. Phillippi: Formation of halos around West Indian patch reefs. Science 182:715–717.

Orth, R. J. and J. Van Montfrans. 1984. Epiphyte-seagrass relationships with an emphasis on the role of micrograzing: A review. Aquat. Bot. 18:43–70.

Ostrofsky, M. L. and E. R. Zettler. 1986. Chemical defences in aquatic plants. J. Ecol. 74:279–287.

Paine, R. T. 1980. Food webs: Linkage, interaction strength and community structure. J. Anim. Ecol. 49:667–686.

Paine, R. T. and S. A. Levin. 1981. Intertidal landscapes: Disturbance and the dynamics of pattern. Ecol. Monogr. 51:145–178.

Pelikan, J., J. Svoboda, and J. Kvet. 1971. Relationship between the population of muskrats (*Ondatra zibethica*) and the primary production of cattail (*Typha latifolia*). Hydrobiologia (Bucharest) 12:177–180.

Phillips, G. L., D. Eminson, and B. Moss. 1978. A mechanism to account for macrophyte decline in progressively eutrophicated freshwaters. Aquat. Bot. 4:103–126.

Pickett, S. T. A. and P. S. White, eds. 1985. The ecology of natural disturbance and patch dynamics. New York: Academic Press. 472 pp.

Pimm, S. L. 1982. Food webs. London: Chapman and Hall.

Post, J. R. and D. Cucin. 1984. Changes in the benthic community of a small precambrian lake following the introduction of yellow perch, *Perca flavescens*. Can. J. Fish Aq. Sci. 41:1496–1501.

Pourriot, R. 1977. Food and feeding habits of Rotifera. Archiv fur Hydrobiologie Beiheft 8:243–260.

Power, M. E. 1987. Predator avoidance by grazing fishes in temperate and tropical streams: Importance of stream depth and prey size. *In*: Predation: Direct and indirect impacts on aquatic communities, ed. W. C. Kerfoot and A. Sih, 333–352. Hanover: University Press of New England.

Prejs, A. 1984. Herbivory by temperate freshwater fishes and its consequences. Environ. Biol. Fish. 10:281–296.

Prentki, R. T., M. S. Adams, S. R. Carpenter, A. Gasith, C. S. Smith, and P. R. Weiler. 1979. The role of submersed weedbeds in internal loading and interception of allochthonous materials in Lake Wingra, Wisconsin, USA. Arch. Hydrobiol. Suppl. 57:221–250.

Price, P. W., M. Westoby, B. Rice, P. R. Atsatt, R. S. Fritz, J. N. Thompson, and K. Mobley. 1986. Parasite mediation in ecological interactions. Ann. Rev. Ecol. Syst. 17:487–506.

Reddy, K. R. and W. H. Patrick. 1984. Nitrogen transformations and loss in flooded soils and sediments. CRC Critical Reviews in Environmental Control 13:273–309.

Reise, K. 1979. Moderate predation on meiofauna by the macrobenthos of the Wadden Sea. Helgolander wiss. Meersunters 32:453–465.

Reynolds, C. S. 1984. Ecology of freshwater algae. Cambridge: Cambridge University Press.

Rich, P. H. and R. G. Wetzel. 1978. Detritus in the lake ecosystem. Am. Nat. 112:57–71.

Riley, E. T. and E. E. Prepas. 1984. Role of internal phosphorus loading in two shallow productive lakes in Alberta, Canada. Can. J. Fish. Aquat. Sci. 41:845–855.

Riley, E. T. and E. E. Prepas. 1985. Comparison of the phosphorus-chlorophyll relationships in mixed and stratified lakes. Can. J. Fish. Aquat. Sci. 42:831–835.

Robbins, J. A. 1982. Stratigraphic and dynamic effects of sediment reworking by Great Lakes zoobenthos. Hydrobiologia 91/92:611–622.

Rogers, K. H. and C. M. Breen. 1983. An investigation of macrophyte, epiphyte and grazer interactions. *In*: Periphyton of freshwater ecosystems, ed. R. G. Wetzel, 217–226. The Hague: Dr. W. Junk Publishers.

Ryding, S. O. 1985. Chemical and microbiological processes as regulators of the exchange of substances between sediments and water in shallow eutrophic lakes. Int. Revue ges. Hydrobiol. 70:657–702.

Sand-Jensen, K., C. Prahl, and H. Stockholm. 1982. Oxygen release from roots of submerged aquatic macrophytes. Oikos 38:349–354.

Sand-Jensen, K. and N. P. Revsbech. 1987. Photosynthesis and light adaptation in epiphyte-macrophyte associations measured by oxygen microelectrodes. Limnol. Oceanogr. 32:452–457.

Shapiro, J. and D. I. Wright. 1984. Lake restoration by biomanipulation: Round Lake, Minnesota, the first two years. Freshwat. Biol. 14:371–383.

Sheldon, S. P. 1984. The effects of herbivory and other factors on species abundance and the diversity of freshwater macrophyte communities. Ph.D. Thesis, University of Minnesota. 117 pp.

Sih, A., P. Crowley, M. McPeek, J. Petranka, and K. Strohmeier. 1985. Predation, competition, and prey communities: A review of field experiments. Am. Rev. Ecol. Syst. 16:269–312.

Smith, C. S. and M. S. Adams. 1986. Phosphorus transfer from sediments by *Myriophyllum spicatum*. Limnol. Oceanogr. 31:1312–1321.

Smith, L. M. and J. A. Kadlec. 1985. Fire and herbivory in a Great Salt Lake marsh. Ecology 66:259–265.

Søndergaard, M. 1983. Heterotrophic utilization and decomposition of extracellular carbon released by the aquatic angiosperm *Littorella uniflora* (L.) Aschers. Aquat. Bot. 16:59–73.

Stanczykowska, A., W. Lawacz, J. Mattice, and K. Lewandowski. 1976. Bivalves as a factor effecting circulation of matter in Lake Mikolajskie (Poland). Limnologica 10:347–352.

Stauffer, R. E. 1985. Nutrient internal cycling and the trophic regulation of Green Lake, Wisconsin. Limnol. Oceanogr. 30:347–363.

Stein, R. A. and J. J. Magnuson. 1976. Behavioral response by crayfish to a fish predator. Ecology 57:751–761.

Sterry, P. R., J. D. Thomas, and R. L. Patience. 1983. Behavioural response of *Biomphalaria glabrata* (Say) to chemical factors from aquatic macrophytes including decaying *Lemna paucicostata* (Hegelm, ex Engelm.) Freshwat. Biol. 13:465–476.

Strayer, D. L., J. J. Cole, G. E. Likens, and D. C. Buso. 1981. Biomass and annual production of the freshwater mussel *Elliptio complanata* in an oligotrophic, softwater lake. Freshwat. Biol. 11:435–440.

Strayer, D. 1986. The size structure of a lacustrine zoobenthic community. Oecologia 69:513–516.

Strayer, D. and G. E. Likens. 1986. An energy budget for the zoobenthos for Mirror Lake, New Hampshire. Ecology 67:303–313.

Strong, A. E. and B. J. Eadie. 1978. Satellite observations of column carbonate precipitations in the Great Lakes. Limnol. Oceanog. 32:877–887.

Sumner, W. T. and C. D. McIntire. 1982. Grazer-periphyton interactions in laboratory streams. Arch. Hydrobiol. 93:135–157.

Tessenow, U. and Y. Baynes. 1978. Experimental effects of *Isoetes lacustris* L. on the distribution of Eh, pH, Fe and Mn in lake sediment. Verh. Int. Verein. Limnol. 20:2358–2362.

Thomas, J. D. 1982. Chemical ecology of the snail hosts of schistosomiasis: snailsnail and snail-plant interactions. Malacologia 22:81–91.

Thorp, J. H. 1986. Two distinct roles for predators in freshwater assemblages. Oikos 47:75–82.

Trimbee, A. M. and G. P.Harris. 1984. Phytoplankton population dynamics of a small reservoir: Use of sedimentation traps to quantify the loss of diatoms and recruitment of summer bloom-forming blue-green algae. J. Plank. Res. 6:897–918.

Twilley, R. R., G. Ejdung, P. Romare and W. M. Kemp. 1986. A comparative study of decomposition, oxygen consumption and nutrient release for selected aquatic plants occurring in an estuarine environment. Oikos 47:190–198.

Twinch, A. J. and R. H. Peters. 1984. Phosphate exchange between littoral sediments and overlying water in an oligotrophic North-temperate lake. Can. J. Fish. Aquat. 41:1609–1617.

Wallace, J. B. and J. O'Hop. 1985. Life on a fast pad: Water lily leaf beetle impact on water lilies. Ecology 66: 1534–1544.

Walter, R. A. 1985. Benthic macroinvertebrates. *In*: An ecosystem approach to aquatic ecology, ed. G. E. Likens, 280–291. New York: Springer-Verlag.

Warwick, R. M. 1984. Species size distributions in marine benthic communities. Oecologia 61:32–41.

Werner, E. E., J. F. Gilliam, D. J. Hall, and G. G. Mittelbach. 1983. An experimental test of the effects of predation risk on habitat use in fish. Ecology 64:1540–1548.

Werner, E. E. and J. F. Gilliam. 1984. The ontogenetic niche and species interactions in size-structured populations. Ann. Rev. Ecol. Syst. 15:393–426.

Wetzel, R. G. 1979. The role of the littoral zone and detritus in lake metabolism. Archiv fur Hydrobiologie 13:145–161.

Wetzel, R. G. 1983a. Attached algal-substrata interactions: fact or myth, and when and how? *In*: Periphyton in Freshwater Ecosystems, ed. R. G. Wetzel, 207–215. The Hague: Dr. W. Junk Publishers.

Wetzel, R. G. 1983b. Limnology, 2d ed. Philadelphia: Saunders College Publishing. 767 pp.

Wium-Anderson, S., H. Anthoni, C. Christophersen, and G. Houen. 1982. Allelopathic effects on phytoplankton by substances isolated from aquatic macrophytes (Charales). Oikos 39:187–190.

Zaret, T. M. and R. T. Paine. 1973. Species introduction in a tropical lake. Science 182:449–55.

CHAPTER 13

Microbial Interactions in Lake Food Webs

K. G. Porter, Chair, H. Paerl, Rapporteur,
R. Hodson, M. Pace, J. Priscu, B. Riemann,
D. Scavia, and J. Stockner

Introduction

Microbial processes are major components of global biogeochemical cycles. Microorganisms play an active role in regulating organic carbon supply and demand, nutrient recycling, and CO_2 and O_2 balances within and between systems. They are involved in fixation and regeneration of the major organic building blocks (carbon, nitrogen, sulfur, and phosphorus) and can mobilize elemental cofactors and vitamins that are essential for the growth and metabolism of all living cells.

The specific roles microorganisms play in freshwater food webs have yet to be determined. We feel that the term microbial loop, coined through work in marine systems (Azam et al. 1983) is potentially misleading in that it implies a system that can be easily separated from, or serves as an appendage to, the classic planktonic food web. Instead, we contend that microbial processes should be considered an integral part of aquatic and other food webs.

Membership and Interactions in the Microbial Component of Planktonic Food Webs

The microbial component of planktonic food webs can be described by taxon, size class, functional group and trophic level with varying degrees of success. Table 13.1 lists the prokaryotic and eukaryotic members of the web as they are known to date. The picoplankton are composed of small autotrophic and heterotrophic cells which provide the major pathway for soluble organic and inorganic nutrients entering the microbial system.

Microbial activity in lakes and other aquatic environments was originally studied by isolation and cultivation of individual species. Traditional laboratory viable-count methods were widely used to enumerate microorganisms. However, once direct microscopic methods of enumeration were developed it became clear that only a small percentage (typically 0.1-1.0%)

TABLE 13.1. Microbial components of the freshwater food web (A = autotrophic, H = heterotrophic, P = phagotrophic)

Microbial component	Trophic mode
Picoplankton (0.2–2 μm)	
bacteria	H
photo & chemolithotrophs	A
coccoid cyanobacteria	A, H
small eukaryotes	A, H
Nannoplankton (2–20 μm)	
flagellates	
Heterotrophic microflagellates	H
Autotrophic microflagellates	A, H
Mixotrophic flagellates	A, H, P
Ciliates	
colorless	P, H
pigmented	P, A, H
Amoebae	P, A, H
Euglenoids	
colorless	P, H
pigmented	A, H
Microzooplankton (20–200 μm)	Omnivore
Ciliates	Omnivore
Rotifers	Omnivore
Nauplii	Omnivore
Macrozooplankton (200 μm–2 mm)	
Cladocera	Omnivore
Copepods	Omnivore

of the bacteria present were cultured on enrichment media. Thus, the earlier enumerations were low by several orders of magnitude. Consequently, previous estimates of carbon flow through bacteria in lakes had suggested no quantitative significance for these organisms. Moreover, laboratory studies with isolates from aquatic systems were suspect for two reasons: growth on a laboratory medium could mean it was atypical (after all, 99.9% of the population wouldn't grow!) and 2) the environmental conditions under which an isolate would grow were recognized to be radically different from natural *in situ* conditions.

Naturally occurring bacteria are now sampled and enumerated by direct epifluorescent counts (the preferred but tedious and imprecise method). Currently they are still generally treated as a black box without regard to taxonomic or functional diversity. Grouping by size class, shape, or in relation to other particulate organic matter, as attached or unattached cells, can be made from direct counts or scanning electron microscopy with considerable labor beyond that expended for the determination of numerical abundance. Differences in function and activity within the bacterial community can only be determined from natural samples by microautoradiography and immunolabeling techniques.

In general, unattached heterotrophic bacterioplankton are in the size range from 0.2-1.0 um in maximum linear dimension although some cells may exceed this range. Coccoid cyano-bacteria and small green algae, which constitute autotrophic as well as heterotrophic components of the picoplankton, are generally 0.5 to 2.0 um when occurring as single cells in fresh waters. This makes separation by size fractionation difficult. No information is available on the relative importance of autotrophy and heterotrophy in their metabolism.

Protists in freshwater planktonic systems include ciliates, amoebae, and flagellates. These have a broad size range, from 2-200 um. Taxonomic distinctions can be made during direct counts; however, functional distinctions are difficult. For example, pigmented ciliates and flagellates may function as primary producers, particle feeders, and consumers of dissolved organic compounds. Size fractionation of these protists from the picoplankton is difficult owing to size overlap as well as to disruptive effects of sieving and centrifugation. Even if these methods are successful, the potential mixotrophic capacity of individual cells makes size or taxonomic classification questionable in determining function *a priori.*

Rotifers range in size from 50-200 um and can be classified primarily as consumers of particulate matter. However, their potential food resources include not only algal and bacterial picoplankton, but also phytoplankton (ranging in size from flagellates to filamentous cyanobacteria), ciliates, nauplii, and other rotifers. Crustacean zooplankton are also potentially omnivorous. Larval fish in their earliest life stages utilize ciliates, rotifers, and nauplii as food sources. Omnivory appears to be the rule rather than the exception for these groups.

Because of the recent focus on the functional role of microbes in aquatic ecosystems, a variety of complex trophic linkages have been discovered (or in some cases rediscovered) within microbial food webs. It is clear that within the microbial component of the planktonic food web, taxonomic and size distinctions do not automatically determine function. Producers in the system convert soluble organic and inorganic nutrients to particulate biomass and have the potential to do so by both autotrophic and heterotrophic metabolic pathways. Consumers of bacterial and algal picoplankton biomass are often omnivorous, ingesting larger algae as well as protists and metazoans. Some are mixotrophic, utilizing particulate and dissolved organic matter as well as functioning as primary producers. Unfortunately, we do not know the relative importance of these trophic interactions. Fig. 13.1 illustrates a current view of the interactions among functional groups within the microbial food web. These complex interactions within a given taxon or cell make *in situ* determination of functional state essential for understanding their roles in a given community. Future research must go beyond identifying the linkages to ascertaining the strength and significance of these predator-prey interactions within microbial communities. In addition to the direct consequences of predatory

COASTAL

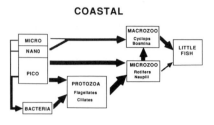

INTERIOR

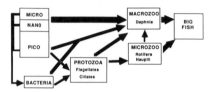

FIGURE 13.1. Predominant pathways of microbial carbon transfer in ultraoligo-trophic coastal lakes and oligotrophic interior lakes in British Columbia. Black arrows denote major fluxes; gray arrows denote minor fluxes. See text for further explanation.

mortality, research should also focus on the potentially important feed-backs in these trophic relations wherein nutrient and dissolved organic carbon released by the predators may stimulate their prey.

Differential Microbial Activity Along Seasonal and Trophic Gradients

Planktonic community structure is the key to how microbial production moves through the food web. An example from the coastal and interior lakes of western Canada (Stockner and Shortreed 1987) illustrates the po-tential importance of *Daphnia* as a keystone species (Fig. 13.1). In ultra-oligotrophic coastal lakes, carbon flow from picoplankton through pro-tozoa, rotifers, and nauplii leads to characteristically small bodied fish populations and low fish production. Oligotrophic interior lakes, in which picoplankton, larger phytoplankton, and protozoan carbon flows directly and indirectly through *Daphnia*, are typically characterized by larger fish and higher fish production. Low levels of fertilization in these lakes reg-ulate these two distinct pathways. High levels of fertilization in Lake Eunice (Neill this volume) transformed a *Daphnia* dominated system into a rotifer-copepod system. This suggests that at least at the oligotrophic end of the trophic continuum, enrichment can have a major effect on de-termining whether the microbial loop is a source or sink of carbon to the community.

Recent evidence suggests that temperature can regulate the coupling and uncoupling of microbial processes from higher trophic levels in aquatic environments (Pomeroy and Diebel 1986). Heterotrophic and autotrophic picoplankton appear to be strongly regulated by temperature (Caron et al. 1985, Scavia and Laird 1987), while nanno- and microplankton communities are less affected (Pomeroy and Diebel 1986). From the limited knowledge of the lake picoplankton, we propose the following model. In dimictic oligotrophic lakes before stratification, large increases in autotrophic production, primarily by microplankton (diatoms), overshadow the temperature regulated picoplankton. During this period, some of this new primary production passes to the traditional, classic consumer web. With the onset of stratification, this new production usually sinks to the hypolimion leaving the epilimnion conditioned with elevated DOM, a paucity of available P and N, and an actively growing zooplankton community. The epilimnion can subsequently be characterized by the dominance of nannoplankton and recycling processes associated with the picoplankton and their predators, which show markedly increased activity at this time (Fig. 13.2B). After stratification, sequestering of DOM and nutrients by the autotrophic and heterotrophic picoplankton, coupled with minimal sedimentation loss, provide a carbon resource for epilimnetic consumers that would be lost in the absence of picoplankton. This condition prevails until the breakdown of stratification, which entrains hypolimnetic nutrients and lowers temperature. New autotrophic production increases again and lower temperature slows microbial activity (MOM; Fig. 13.2A). During winter mixing autotrophic production is light-limited and, while microbial process rates are lower, the persistent slow mineralization of DOM partially resets conditions for spring autotrophic production. Thus, when considering the seasonality of traditional autotrophic and microbial processes, the relative importance of microbial components such as carbon and inorganic nutrient sources is likely confined to the summer epilimnion.

If microbial carbon production is significant during summer stratification, then that carbon is provided to the epilimnetic zooplankton community at a critical time when autotrophic microplankton production is most nutrient stressed. While the presence of *Daphnia* apparently leads to more efficient transfer of carbon from the microbial food web (see below), the relative significance of that carbon flux as compared to that from the nano- and microplankton is currently unknown. In warm monomictic coastal lakes, characterized by minimal external supplies of substrate (DOM) and inorganic nutrients, this window of microbial activity is protracted owing to the absence of any significant spring pulse of new production (Stockner and Shortreed 1985, Stockner 1987). At the other end of the trophic spectrum are eutrophic lakes where allochthonous nutrient loading can persist throughout summer. New production can be sustained over the season in these lakes, and autotrophic picoplankton production is a smaller and less significant fraction of total carbon production (Stock-

MEASURE OF MICROBIAL STRENGTH

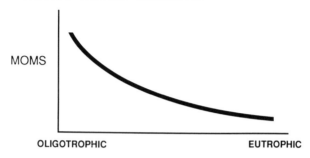

$$\text{MOMS} = \frac{\text{Autotrophic Picoplankton} + \text{Bacterial Production}}{\text{Total Autotrophic} + \text{Heterotrophic Production}}$$

A. ACROSS TROPHIC GRADIENTS

MOMS

OLIGOTROPHIC   EUTROPHIC

B. SEASONALLY WITHIN A TEMPORATE DIMICTIC LAKE

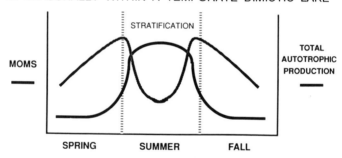

STRATIFICATION

TOTAL
AUTOTROPHIC
PRODUCTION

MOMS

SPRING   SUMMER   FALL

FIGURE 13.2. The importance of microbial production in lake ecosystems in relation to *(A)* trophic state and *(B)* season in a temperate dimictic lake.

ner and Shortreed 1987). In general, the significance of microbial processes increases along a trophic gradient leading to oligotrophy, where its importance is heightened in the summer epilimnion during the period of stratification (Fig. 13.2B).

Implication of Complex Microbial Interactions for the Transfer of Material and Energy to the Classic Web

There has been considerable recent debate regarding whether a significant percentage of the carbon entering the microbial food web is transferred to higher trophic levels and how important such pathways are in the trophodynamics of the food webs. Two points need to be considered in such a discussion. First, no trophic transfer is 100% efficient. Thus, each added step in carbon flow is, in fact a sink for carbon, as some will be lost to respiration. However, kinetics are more directly at issue here than are thermodynamics. Second, microbial loop components can be considered links only to the extent that carbon is recovered which would otherwise be lost to the animal food web for a long period on an ecosystem time scale. For example, carbon that is stored as unavailable geopolymer for a period equal to the lifetime of the lake, or for a period sufficiently long for the carbon to be advectively removed from the system, is effectively lost to higher trophic levels.

A key question concerning the transfer of material from the microbial web is the nature of the interface with the classic web (Fig. 13.1). The interface is controlled largely by the feeding characteristics of the consumers in the classic web. In lakes with few *Daphnia*, rotifers may play a significant role in bridging the microbial and classic webs. Rotifers graze components of the microbial web (Bogdan and Gilbert 1982, Stockner 1987), and are in turn consumed by copepods (Williamson and Butler 1986). However, these additional trophic steps will lead to relatively low transfer efficiency to higher trophic levels. *Daphnia* appears to be a keystone species (Stockner and Porter, this volume) since it consumes prey from a size spectrum encompassing several trophic levels: heterotrophic and autotrophic picoplankton, nannoplankton, and microplankton (Porter 1977). In contrast, other zooplankton (e.g., copepods) are restricted to grazing only larger prey (Vanderploeg et al. 1984). Thus, the presence of *Daphnia* markedly shortens the food chain and increases transfer efficiency (Stockner and Porter this volume).

Because pelagic fish communities can strongly influence the size structure and species composition of the zooplankton community (Brooks and Dodson 1965, Scavia et al. 1987), the nature and efficiency of energy transfer from microbial to classic food webs can be driven by interactions high in the trophic structure. For example, intense zooplanktivory reduced the significance of large-bodied *Daphnia* and thus lowered the efficiency of carbon transfer from the microbial food web in large-scale enclosure manipulations (Riemann 1985).

Spatial/Temporal Scales: Impacts on Microbial Production Rates and Transfer Characteristics in Food Webs

The characterization, regulation, and trophic impacts of microbial production and nutrient cycling are intimately tied to certain physical and temporal scales in freshwater ecosystems. Within microbially mediated food webs, metabolic coupling and trophic transfer may be extremely tight, such as in (1) bacterial, microflagellate, and protozoan associations with phycospheres of algae (2) chemotactically-mediated grazing and (3) parasitism or commensalism. The coupling may be relatively loose in cases where production, grazing, and consumption are mediated over large (mm-m) linear gradients in the water column or over seasonal time scales.

Microbially important spatial scales range from submicrons to meters. There exist numerous instances where microbial production and cycling processes exhibit specific small- or large-scale requirements. For example, primary producers orient and maintain populations in specific photosynthetically active radiation (PAR) regimes on either a semipermanent basis (e.g., surface and metalimnetic cyanobacterial populations in stratified systems) or along a fluctuating gradient (as is the case among numerous procaryotic and eucaryotic picoplankton and microplankton species in mixed waters). These orientations are critical spatial requirements for optimizing daily production rates and growth. A variety of investigators (Harris 1978; Platt and Gallegos 1980; Falkowski 1983) have concluded that unless natural microbial communities are exposed to appropriate light/ nutrient regimes (either highly variable or confined to specific depths along vertical gradients) with respect to environmental requirements, it is unlikely that realistic assessments of production rates will result. In terms of metabolic coupling and matter/energy transfer to higher levels of the food web, consumers must specifically be adapted to variable or fluctuating vertical scales. In the case of consumers of surface dwelling primary producers or heterotrophs, adaptations must include the ability to counteract potentially negative impacts of UV irradiance, dramatic temperature fluctations, periodic dessication (in mats) and tremendous O_2, pH, and CO_2 fluctuations. Conversely, grazers of deep-living metalimnetic or hypolimnetic microbial producers may well need to cope with periodic anoxia, cold and saline conditions. Among consumers of microbes thriving under highly mixed conditions, adequate reception (chemosensing, physical sensing, vision) and capture mechanisms are of paramount importance.

Among oxygenic microbial producers and consumers in mixed water columns, spatial confinement appears quite variable and encompasses scales of the order of millimeters to meters. In contrast, O_2 sensitive (anaerobic, microaerophilic) producers and nutrient cyclers must be confined to highly reduced (O_2-devoid) microenvironments (microzones) in systems having high ambient O_2 levels. Microzones such as those present in suspended microbial aggregates and detritus can contain internal O_2 gradients having micron dimensions. Within such gradients, a microbial food web, consisting of primary producers, microheterotrophs, and predators participates in trophic interactions before transfer into more conventional food webs.

This heterogeneity is also reflected in the kinetic response of the microflora with regard to substrate uptake (and hence, the biogeochemistry of the water column). Studies are still routinely published in which the kinetics of substrate uptake from lakes is assumed to be monophasic. Often, the assumption is made that all the microorganisms present are taking up the compound with the same kinetic parameters (Kt; Vmax). This assumption derives from an earlier one that concentrations of dissolved organic compounds are temporally and spatially homogeneous. Since DOC concentrations are not homogeneous, it is unwarranted to assume that there would be selective pressure for microbes to evolve the same, very low Kt value is also not warranted. In fact, the kinetic response of bacterioplankton shows a functional plasticity regarding uptake rate dependence on substrate concentration which has been termed kinetic diversity (Azam and Hodson 1981; Murray and Hodson 1984). The effect is that substrate uptake rates do not saturate and the microbial assemblages immediately respond to temporal and spatial changes in organic carbon availability in a way that maximizes uptake. Thus, uptake kinetics are multiphasic. Prediction of the importance of the microbial component of food chains and community structure (DOC-bacteria-bacterivores-etc.) must consider the real nature of the dependence of bacterial uptake (and production) on substrate concentration.

Both identification of relevant scales and development of methods for assessing organic carbon and inorganic nutrient transfer at such scales are current research challenges. Given recently developed techniques, we have only begun to quantitatively determine the diverse spatial array of links and sinks in planktonic (let alone benthic; Lodge et al., this volume) aquatic environments. Small-scale microzones represent the most formidable technological (as well as conceptual) barriers. However, microautoradiography, immunochemical labeling (for specific processes as well as taxa), and microelectrode techniques have opened avenues for quantifying the trophic roles which microzone-scale consortia play in fresh water and marine ecosystems.

Microbial production, cycling, and trophic transfer processes are also heavily reliant on temporal fluctuations (or conversely, consistencies) among physico-chemical and biotic parameters. The temporal scales of these fluctuations and their daily and seasonal transitions are crucial in regulating rates of primary and secondary production and nutrient transformation. These processes are heavily reliant on irradiance patterns and histories. Among pico-, nanno-, and microplanktonic producers, diel rates of primary production often do not follow quantities of PAR flux (Eppley and Sharp 1975, Paerl and Mackenzie 1977, Harris and Piccinin 1977, Platt and Gallegos 1980). A strong tendency to maximize primary production rates during early to midmorning hours, prior to maximum PAR flux periods, has been noted among phytoplankton. Unfortunately, the timing of maximum photosynthesis can vary from day to day, season to season, and from lake to lake, making predictive modeling based on fixed time scales exceedingly difficult. Furthermore, diel photosynthetic and nutritional histories of phytoplankton communities appear to play a role in determining the nonparallel behavior of primary production patterns and maxima with respect to PAR flux patterns.

The lack of parallelism between photon flux and productivity is of particular relevance when considering the timing and duration of incubation periods required to quantify production. In this regard, standard mid-day 3-4 h incubation periods are not yielding realistic or quantitatively sound measurements of pico- and microplankton primary production. Temporal lags of bacterial production coupled in varying spatial degrees to primary production further complicate extrapolation of temporal coupling based on short-term incubations. Clearly, long-term examinations of diel primary production in relation to heterotrophic utilization within microbial loops are needed.

Diel studies will also be required to obtain ecologically relevant nutrient cycling rates. An example of a nutrient cycling process whose rates often deviate significantly from either length or intensities of irradiance is N_2 fixation, despite the fact that this process is often mediated by photoautotrophs (cyanobacteria). Recent diel studies have shown that under certain circumstances, nighttime N_2 fixation rates can exceed daylight rates in some planktonic and benthic habitats (Bebout pers. comm.). Most likely, high O_2 is a potent inhibitor of N_2 fixation among both cyanobacteria and eubacteria. It follows that it will be necessary to greatly expand the time scale for N_2 fixation measurements in order to realistically assess true ecosystem N inputs (budgets) via this process.

The expansion of time intervals (diel studies) required for quantification of microbial production, cycling and trophic transfer processes necessitates careful consideration of realistic incubation container sizes, in order to avoid potential artifacts induced by long-term bottle effects. Accordingly, careful consideration of container sizes, nutrient availability and optical properties must accompany evaluations of techniques.

Complexities of Microbial Processing in Food Webs

The question as to whether microbial/detrital processes are consequences or determinants of community structure has no single answer; these processes are, in fact, both. For example, one could argue that the plant community structure of a lake is a major determinant of microbial/detrital processes since the quality of organic matter produced (both DOC and POC) will poise the system with regard to the overall types and concentrations of microorganisms that occupy the sediments and water column. For instance, littoral areas and shallow lakes dominated by emergent higher vegetation will have a rich microflora adapted to lignocellulose degradation, whereas open water, phytoplankton dominated habitats will not. In turn, the quality, cell density, and spatial distribution (e.g. percent of cells that are free-living vs attached to detrital particles) of the microbial assemblages will determine how much and what overall percentage of photosynthetically fixed carbon is passed from the procaryotes involved in organic matter dissimilation to higher trophic levels.

To a certain degree, microbial degradative populations can change to reflect changes in the quality of available organic compounds. However, real differences in functional diversity of heterotrophic bacteria exist from habitat to habitat—differences that persist for considerable periods after the quality of organic substrates has been experimentally altered. Hodson and coworkers (in prep.) find, for instance, that the lignocellulose degradation rates supported by the natural mixed microflora of the Okefenokee swamp and other freshwater or marine environments differ. These differences are retained even weeks after the environmental conditions (e.g. pH, nutrients, type of available lignocellulose) in one system have been perturbed to mimic those of the others.

Most of the discussions about the microbial components as links or sinks have centered on pelagic systems. Very little argument has centered on so-called detritus-based systems, such as wetlands, littoral zones of lakes, or shallow lakes with abundant and (to animals) indigestible vascular plant growth. Perhaps this is because in the past so little attention has been paid to the role of bacteria in open water systems, while microbial processing has always been recognized as being important in detritus-based systems. In fact, the operational definition of a detritus-based aquatic ecosystem (Odum and de La Cruz 1967) has been that, in these systems, microbial transformation of indigestible plant matter to microbial biomass and assimilable degradation products serves as the principal link between primary and secondary production. This is the archtypal definition of microbes serving as links.

In point of fact, there is not much information on either the efficiency of carbon transfer or the overall rates of transfer of refractory carbon in these purportedly detritus-based systems. The possibility remains that in some of these systems (marshes, swamps, shallow lakes, etc.), grazing food chains based on photosynthetic carbon fixation by periphyton or phytoplankton could be more important to animal production than food chains based on microbial conversion of detrital organic carbon derived from vascular plants, notwithstanding the abundance of this carbon source. It has been shown by experiment and modeling (M. A. Moran, T. Legovic, R. Benner and R. E. Hodson, unpublished) that only one to a few percent of the carbon stored as newly formed detrital lignocellulose (nascent peat) in the Okefenokee Swamp is incorporated into biomass of detritivorous animals. The bulk is lost to respiration by bacteria, fungi and protozoa. Although the lignocellulolytic microorganisms of the system assimilate carbon with reasonably high efficiency (approx. 30%), this carbon flow is sufficient to account for only an average of 17% of the total production of bacteria in the swamp, suggesting that other carbon sources are also important.

There is a need to reevaluate definitions of detritus-based and link/sink as applied to aquatic systems in which the role of microbial detritus processing is recognized.

Marine Vs. Freshwater Systems

Most recent knowledge of aquatic microbes has arisen from developments in marine environments (Klug and Reddy 1984; Hobbie and Williams 1984), despite the fact that many of the original ideas were derived from freshwater studies (Stockner and Porter this volume; Stockner and Antia 1986). For example, the link/sink debate reviewed above has centered on marine systems (Fuhrman and McManus 1984; Ducklow et al. 1986). Microbial ecological studies in freshwater systems have focused generally on the classic role of microbes as decomposers in major biogeochemical cycles (Kuznetsov et al. 1979). Consequently, freshwater community ecologists have only recently begun to evaluate the trophic importance of the microbial component of the food web (Bell et al. 1983; Bell and Kuparinen 1984; Lovell and Konopka 1985a, b; Scavia et al. 1986; Scavia and Laird 1987).

Freshwater systems offer several advantages for answering questions regarding the importance and regulation of microbial processes in aquatic systems. An obvious advantage is the relative ease and cost-effectiveness with which freshwater systems can be sampled. As a result, detailed seasonal studies of the importance of microbial processes can be undertaken. Freshwater systems also offer a wide range of conditions (e.g. trophic state, temperature, age, morphometry, latitude) across which the role of microbial processes can be evaluated. Moreover, lakes more readily lend themselves to experimental manipulation, either through mesocosm studies or whole-lake perturbations, as compared to the smaller bottle experiments and laboratory cultures required for most marine studies. Finally, material balances can be determined more precisely in freshwater systems (Priscu et al. 1982). Of particular importance in this respect are man made reservoirs. These systems can be viewed as chemostats in which materials (e.g., dissolved inorganic and organic nutrients) are continually being replenished. Because material balances can be more easily constructed in these systems, microbial concepts developed in marine systems, such as new vs regenerated production (Dugdale and Goering 1967; Eppley and Peterson 1979) and the role of carbon transfer through the microbial loop as a link to higher trophic levels (Azam et al. 1983), can be tested.

There is currently a need for synthesis of freshwater and marine studies to obtain global perspectives on the role of microbial communities in aquatic systems. We currently feel that no functional differences exist in the ecological relationships between marine and freshwater microbial systems particularly with regard to inorganic nutrient dynamics (Axler et al.

1981; Priscu and Priscu 1984). Interdisciplinary work is the only way in which to test this contention and to advance our knowledge regarding the importance of the microbial community in regulating interactions in aquatic systems.

Experiments, Methodological Problems, and the Need for Novel Techniques

Microbial food webs are clearly not amenable to traditional manipulations, since it is difficult to either selectively remove or add microorganisms to the community. Instead, microbial ecologists must develop techniques for detecting links and measuring the rates of predation, nutrient regeneration, and dissolved organic carbon release by functional groups of the microbial food web. Application of such techniques across trophic gradients should help to identify what pathways are important and in which systems they are important. Additionally, experiments wherein the dynamics of the community are followed after an induced perturbation (e.g. nutrient pulsing) should help elucidate important interactions. Coupling these experiments with a mathematical model of the system might also suggest what key factors are controlling the community components. Experimental approaches of this nature would help move the subject from a descriptive phase to an understanding of how the functional groups within microbial communities control the overall movement of material and energy through these communities. We are currently limited by our ability to assess the rate processes within these communities and our poor understanding of their dynamics on the most relevant time scales of hours to days.

In our attempts to understand the flux of matter between dissolved pools, microorganisms, and higher trophic levels, it is important to measure biomasses as well as rates of production. A number of conversion factors are used to relate cell numbers and various rate estimates to carbon biomass and production. In view of ongoing discussions in the literature concerning the range of these factors, the choice to combine them is debatable. Because ecological conclusions from field data are highly influenced by the set of conversion factors applied, it is recommended that more attention be focused on developing proper conversion factors. Parallel consideration should be given to both microbial and higher trophic transformation.

As an example of interacting conversion factors, four conversion factors are needed before measurements of ^{3}H-thymidine incorporation rates can be converted into bacterial carbon gross production. A typical literature-based procedure includes (1) a conversion factor of about 2×10^{18} cells per mole thymidine incorporated into bacterial DNA, (2) a mean cell volume of 0.01-0.3 um^3, (3) a carbon content of 1.121 pg C um^{-3}, and (4) a growth efficiency (growth/uptake) of 60%. In practice, the mean cell volume can be determined microscopically, but for the remaining three, literature-based conversion factors are usually applied. A number of recent studies have suggested that conversions factors should be lowered to 1.1 x 10^{18}, at least in coastal marine environments (Riemann et al. 1987). However, studies on Lake Michigan suggest that not only should the factor be raised (2-20 x 10^{18} cells produced per mole thymidine incorporated in cold TCA fractions), but that it varies seasonally. More recent studies suggest that the carbon content of bacteria should be raised from 1.121 pg C um^{-3} to 0.350 pg C um^{-3} (Bratbak 1985, Bjornsen 1986a) and the growth efficiency should be lowered to 25% (Bell and Kuparinen 1984, Bjornsen 1986b). Applying the new lower factor (#1) and higher factors (#3) and (#4) would suggest that values of both bacterial biomass and carbon gross production should be raised by a factor of 3 compared with results obtained from the use of the conventional conversion factors. Applications of these new factors leads to new interpretations of the role of bacteria in the aquatic environments.

Presently, aquatic environments face new, potentially severe perturbations following the release and proliferation of novel, genetically altered microorganisms. There is wide agreement that it will be critical to follow such organisms as they disperse through the environment to determine how long they persist and what their effects might be on preexisting populations in the water and sediments and on the surfaces of submersed plants. To look for potential displacement by the introduced organism of the natural microflora will require that we first characterize the natural microflora with regard to species (or genetic) diversity and relative abundances as well as growth rates. All of these goals require us to track individual organisms.

At an even more highly resolved level of discrimination, it will be necessary to track the fate of individual genes introduced into the environment using the new species as a vehicle. One such scenario could be that an isolate is taken from a natural aquatic environment and brought into an industrial lab, where it is altered to differ from the original at only one gene locus or a few gene loci. Then it might be released into the environment again in hopes of altering some system parameter (e.g., it might be altered to enhance its ability to degrade some toxic organic compound). To find the introduced organism in nature we will have to be able to discriminate at the gene, rather than at the species, level. To further complicate the picture, if the introduced gene is located on a plasmid or transposon, the host organism might donate its new gene to other, naturally occurring bacteria in the ecosystem. To trace the fate of this introduced genetic material, it would have to be followed from donor to recipient.

All of these possibilities require that individual species, and more specifically, individual DNA sequences (one or more genes) be traceable in natural populations of aquatic bacteria, notwithstanding the fact that many or most of the organisms are not (yet) culturable. This need puts a high premium on development of techniques to relocate a strain or its associated DNA without having to first grow the organism. Such techniques exist in the laboratories of microbial geneticists, but require extensive modification and validation if they are to be applicable to natural samples.

Monoclonal and polyclonal antibody techniques can be used to follow the cell envelopes of introduced organisms or natural organisms that have been previously cultured. Pending development of appropriate permeation steps, antibodies can also be effective in identifying and quantifying specific enzymes and other intracellular constituents involved in production and nutrient transformation processes (e.g., nitrogenase among N2 fixing microorganisms; RUBP-carboxylase among photosynthetic autotrophs). DNA/DNA hybridization techniques are potential tracers of individual genes, provided that several copies of the gene are present in each organism and that sufficiently sensitive procedures (fluorescence, isotopic tracers) for visualizing the bound DNA can be developed. All these techniques, when developed, will bot assist in studies of potential ecosystem perturbation by genetically engineered microorganisms and present real opportunities for basic studies of the population ecology and physiological functions (roles) of aquatic bacteria-studies that in the past have been technically impossible.

These techniques point to another technological need. All the methods mentioned above will require literally tens of thousands of microbial enumerations-far too many to be effectively and precisely done by technicians looking through the microscope. Thus, other important and versatile methodologies that will greatly advance freshwater microbiology studies will be automated counting, sizing and characterization procedures. Current equipment for flow cytometry and image analysis could be used to automate most such counting procedures if the proper software were developed and bundled with the hardware. Moreover, the use of multiple filters (real and software/virtual filters) to view the samples would enable many parameters to be followed simultaneously-an important consideration for samples with short usable lifetimes.

Complex Interactions in Aquatic Communities: Microbial Communities as Model Systems

Many of the theoretical issues in community ecology require appropriate model systems for experimental tests. There is a long history in ecology of using microcosms for this purpose. With the increased knowledge and sophisticated methods available in the field of microbial ecology, there appears to be an opportunity to use complex microbial communities to test general ideas. For example, the concepts of top-down vs bottom-up control have recently received considerable attention in aquatic ecology (Crowder et al. this volume). Experimental and theoretical studies have demonstrated that variance in the recruitment of piscivores may have a strong impact on the organization and variation of aquatic communities through time (Carpenter this volume). The best examples of these processes come from studies of fish; however, direct testing of these ideas requires whole-lake fish manipulations and/or long-term studies of dynamics. Microbial communities should reproduce many of the same patterns over very fast time scales in relation to the life span of the investigator. It should be feasible to manipulate the recruitment of the top predator of a pelagic microbial community and observe how this affects the nature of the community through many generations. Furthermore, it should be possible to design appropriately replicated experiments with microbial communities which would overcome many of the recent objections to experiments performed in community ecology. We thus recognize an undeveloped opportunity to utilize microbial communities to examine general theories within the field of community ecology and recommend increased interactions among theoreticians, experimental community ecologists, and microbial ecologists.

Summary

The following are areas of immediate research need:

1. We need to know the most basic information concerning the pathways and rates of transfer among members of the microbial community and other components of the plankton. The range of metabolic possibilities can be elucidated in laboratory cultures and microcosms but, ultimately, rates and relationships must be determined *in situ*. As a first step we recommend determining broad seasonal states and intense diel studies within each state over an annual cycle.
2. The small size of members of the microbial community does not necessarily result in interactions that occur on small spatial and temporal scales. Although doubling and recycling rates may be at the level of minutes and submicrons, indirect effects may occur on the scales of years and meters.
3. Cross-system comparisons of microbial activity can help determine the factors coupling microbial production to higher trophic levels. Seasonal and spatial patterns have been determined within only a few lakes to date. Information along the oligotrophic-eutrophic spectrum of lakes can be used to predict conditions that regulate linkage intensity. Latitudinal gradients from cold dimictic and warm monomictic and polymictic systems are worth examining in this regard. Marine and freshwater comparisons, already common to microbial ecologists, need to be extended at the community level.
4. Top-down regulation of microbial and biogeochemical processes can

be assessed by including the microbial community in sampling procedures for standard fish and zooplankton manipulation experiments.

5. Study of bottom-up control of higher trophic levels is limited by difficulties in caging or removing the microbes. However enhancement of microbial activity by graded organic enrichment is a possibility.

6. Basic information on functional response (e.g., determination of threshholds in apparent limiting concentrations, etc.) of bacteriovores has yet to be determined. Selective grazing on the basis of size, taste, or digestibility is also unknown.

7. Physical/chemical control of the microbial community must ultimately be studied *in situ*.

8. The nature and function of the anaerobic microbial community found in the anoxic hypolimnion of most lakes is virtually unknown.

9. Direct utilization of detritus as a food resource by higher trophic levels is also poorly known.

10. Study of the fate and impact of bioengineered microorganisms in natural microbial communities is of both theoretical and practical importance.

11. Microbes have been, and will continue to be used as models to test theory. Microbial microcosms and mesocosms have replicability, short duration, and short time lags that make them ideal systems for some sorts of studies.

12. Techniques are within our grasp for the study of the microbial component of planktonic food webs. Automated counting techniques (e.g., image analysis and flow cytometry) and methods for determining identity and metabolic state (i.e., immunofluorescent, immunolocalization and DNA probes) can be adapted from cell and molecular biology to meet our needs. Without them, the task of determining spatial and temporal patterns is insurmountable.

13. Better physiological conversion factors for assessing heterotrophic bacterial production are needed. It is particularly important to evaluate the variation as well as magnitudes of these factors.

References

Azam, F., T. Fenchel, J. G. Field, J. S. Grey, L. A. Meyer-Reil and F. Thiungstad. 1983. The ecological role of water-column microbes in the sea. Mar. Ecol. Prog. Ser. 10:257–263.

Azam, F. and R. Hodson. 1981. Multiphasic kinetics for D-glucose uptake by assemblages of natural marine bacteria. Mar. Ecol. Prog. Ser. 6:213–222.

Axler, R. P., G. W. Redfield and C. R. Goldman. 1981. The importance of regenerated nitrogen to phytoplankton productivity in a subalpine lake. Ecology 62:345–354.

Bell, R. T., G. M. Ahlgren and I. Ahlgren. 1983. Estimating bacterioplankton production by measuring [³H]thymidine incorporation in aeutrophic Swedish Lake. Appl. Environ. Microbiol. 45:1709–1721.

Bell, R. T. and J. Kuparinen. 1984. Assessing phytoplankton and bacterioplankton production during early spring in Lake Erken, Sweden. Appl. Environ. Microbiol. 48:1221–1230.

Bjornsen, P. K. 1986a. Automatic determinations of bacterioplankton biomass by image analysis. Appl. Environ. Microbiol. 51:1199–1204.

Bjornsen, P. K. 1986b. Bacterioplankton growth yield in continuous seawater cultures. Mar. Ecol. Prog. Ser. 30: 191–196.

Bogdan, K. G. and J. J.Gilbert. 1982. Seasonal patterns of feeding by natural populations of *Keratella, Polyarthra* and *Bosmina*: Clearance rates, selectivities, and contributions to community grazing. Limnol. Oceanogr. 27:918–934.

Bratbak, G. 1985. Bacterial biovolume and biomass estimations. Appl. Environ. Microbiol. 49:1488–1493.

Brooks, J. L. and S. I. Dodson. 1965. Predation, body size, and composition of the plankton. Science 150:28–35.

Caron, D. A., F. R. Pick and D. R. S. Lean. 1985. Chroococcoid cyanobacteria in Lake Ontario: Vertical and seasonal distribution during 1982. J. Phycol. 21:171–175.

Ducklow, H. W., D. A. Purdie, P. J. LeB. Williams and J. M. Davies. 1986. Bacterioplankton: A sink for carbon in a coastal marine plankton community. Science 232: 865–867.

Dugdale, R. C. and J. J. Goering. 1967. Uptake of new and regenerated forms of nitrogen productivity. Limnol. Oceanogr. 12:196–206.

Eppley, R. W. and B. J. Peterson. 1979. Particulate organic matter flux and planktonic new production in the deep ocean. Nature 282:677–681.

Eppley, R. W. and J. H. Sharp. 1975. Photosynthetic measurements in the central North Pacific: The dark loss of carbon on 24–h incubations. Limnol. Oceanogr. 20:981-987.

Falkowski, P. G. 1983. Light shade adaptation and vertical mixing of marine phytoplankton: A comparative field study. J. Mar. Res. 41:215–237.

Fuhrman, J. A. and G. B. McManus. 1984. Do bacteria-sized marine eukaryotes consume significant bacterial production? Science 224:1257–1259.

Harris, G. P. 1978. Photosynthesis, productivity and growth: The physiological ecology of phytoplankton. Arch. Hydrobiol. Beih. Ergehn. Limnol. 10:1–171.

Harris, G. P. and B. B. Piccinin. 1977. Photosynthesis by natural phytoplankton populations. Arch. Hydrobiol. 80: 405–457.

Hobbie, J. E. and P. J. LeB. Williams, eds. 1984. Heterotrophic activity in the sea. New York: Plenum.

Klug, M. J. and C. A. Reddy, eds. 1984. Current perspectives in Microbial Ecology. Am. Soc. Microbiol. Washington, D.C.

Kuznetsov, S. I., G. A. Dubinina and N. A. Lapteva. 1979. Biology of oligotrophic bacteria. Ann. Rev. Microbiol. 33:377–387.

Lovell, C. R. and A. Konopka. 1985a. Primary and bacterial production in two dimictic Indiana Lakes. Appl. Environ. Microbiol. 49:485–491.

Lovell, C. R. and A. Konopka. 1985b. Seasonal bacterial production in a dimictic lake as measured by increases in cell numbers and thymidine incorporation. Appl. Environ. Microbiol. 49:492–500.

Murray, R. E. and R. E. Hodson. 1984. Microbial biomass and utilization of dissolved organic matter in the Okeefenokee Swamp ecosystem. Appl. Environ. Microbiol. 47:685–692.

Odum, E. P. and A. A. de la Cruz. 1967. Particulate organic detritus in a Georgia salt marsh—estuarine ecosystem. in: Estuaries, ed. G. H. Lauff. Amer. Assoc. Adv. Sci. Publ. 83: 383–388.

Paerl, H. W. and A. L. Mackenzie. 1977. A comparative study of the diurnal carbon fixation patterns of nannoplankton and net plankton. Limnol. Oceanogr. 22:732–738.

Platt, T. and L. L. Gallegos. 1980. Modelling primary production. in: Primary production in the sea, ed. P. G. Falkowski, 339–362. New York: Plenum Press.

Pomeroy, L. R., and D. Deibel. 1986. Temperature regulation of bacterial activity during the spring bloom in Newfoundland coastal waters. Science 233:359–361.

Porter, K. G. 1977. The plant-animal interface in freshwater ecosystems. Amer. Sci. 65:159–170.

Priscu, J. C. and L. R. Priscu. 1984. Inorganic nitrogen uptake in oligotrophic Lake Taupo, New Zealand. Can. J. Fish. Aquat. Sci. 41:1436–1445.

Priscu, J. C., J. Verduin and J. E. Deacon. 1982. Primary productivity and nutrient balance in a lower Colorado River Reservoir. Arch. Hydrobiol. 94:1–23.

Riemann, B. 1985. Potential influence of fish predation and zooplankton grazing on natural populations of freshwater bacteria. Appl. Environ. Microbiol. 50:187–193.

Riemann, B., P. K. Bjornsen, S. Y. Newell and R. Fallon. 1987. Calculation of cell production of coastal marine bacteria based on measured incorporation of ³H-thymidine. Limnol. Oceanogr. 32:471–476.

Scavia, D. and G. A. Laird. 1987. Bacterioplankton in Lake Michigan: Dynamics, controls, and significance to carbon flux. Limnol. Oceanogr. 32: in press.

Scavia, D., G.A. Laird and G.L. Fahnenstiel. 1986. Production of planktonic bacteria in Lake Michigan. Limnol. Oceanogr. 31: 612–626.

Scavia, D., G. A. Lang and J. F. Kitchell. 1987. Dynamics of Lake Michigan plankton: A model evaluation of nutrient loading, competition, and predation. Can. J. Fish. Aquat. Sci., in press.

Stockner, J. G. 1987. Lake fertilization: the enrichment cycle and lake sockeye salmon production. Can. Spec. Publ. Fish. Aquat. Sci.: in press.

Stockner, J. G. and N. J. Antia. 1986. Algal picoplankton from marine and freshwater ecosystems: a multidisciplinary perspective. Can. J. Fish. Aquat. Sci. 43: 2472–2503.

Stockner, J. G. and K. S. Shortreed. 1985. Whole-lake fertilization experiments in coastal British Columbia lakes: empirical relationships between nutrient inputs and phytoplankton biomass and production. Can. J. Fish. Aquat. Sci. 42: 649–658.

Stockner, J. G. and K. S. Shortreed. 1987. Algal picoplankton production and contribution to food webs in oligotrophic British Columbia lakes. Hydrobiologia: in press.

Vanderploeg, H. A., D. Scavia and J. R. Liebig. 1984. Feeding rate of Diaptomus sicilis and its relation to selectivity and effective food concentration in algal mixtures and in Lake Michigan. J. Plankton Res. 6:919–941.

Williamson, C. E. and N. M. Butler. 1986. Predation on rotifers by the suspension-feeding calanoid copepod Diaptomus pallidus. Limnol. Oceanogr. 31:393–402.

CHAPTER 14

Scale in the Design and Interpretation of Aquatic Community Research

Thomas M. Frost, Donald L. DeAngelis,
Steven M. Bartell, Donald J. Hall,
and Stuart H. Hurlbert

Introduction

The scales employed in investigations of aquatic ecosystems can strongly influence interpretations of community patterns and processes. Some examples are obvious; in contrasting cladocerans and rotifers, assessments of biomass (an instantaneous time scale) provide strikingly different impressions than assessments of production (a broader time scale) (Makarewicz and Likens 1975). Other examples are more subtle but equally important; seasonal distributions of phytoplankton species over several years may indicate patterns with little predictability but distributions of functional groups of phytoplankton can indicate a periodic behavior (Bartell et al. 1978; Reynolds 1984). Explicit considerations of the scales that an observer uses in an investigation are fundamental to an understanding of aquatic systems.

The complexity of structure and function in aquatic communities makes scale considerations particularly important. Allen and Starr (1982) have suggested that the predictability of a system's behavior is a function of the number of its specified components (see also Harris 1986; O'Neill et al. 1986). They describe small number systems in which the behavior of relatively few components may be predicted by models that specifically identify all parts (their examples include the behavior of billiard balls and planets). On the opposite extreme, when systems involve large numbers of components, the behavior of individual components cannot be predicted but the general tendencies of the whole system can be projected based on the average behavior of system components (e.g., gas laws). Between these extremes are middle number systems with behavior that is too complex to predict individually but with too few parts to focus only on general statistical properties (that a bubble will form in a freshly poured glass of champagne is predictable, but when and where it will form is not). Allen and Starr (1982) ascribe middle number character to most specifications of ecological systems, but within the field of ecology itself, their examples suggest that it is community questions which are most strongly influenced by middle number problems.

Within physiological and population ecology, equilibrium models can be fairly successful at predicting system behavior. A system with two or three species under carefully controlled environmental conditions will often be highly predictable and describable by simple sets of differential equations (Pimm 1982). On the other extreme, models of ecosystems that deal with large-scale processes within a system (such as net primary production), i.e., models integrating large numbers of components, can also be fairly successful at predicting overall system behavior (Vollenweider 1968). Community studies that attempt to predict the behavior of particular system components in their interaction with numerous other system parts fall within the middle number realm.

Having recognized the middle number nature of their questions, community ecologists are faced with two problems. They must choose appropriate components or sets of components for which to examine behavior, a problem of theory. They also must choose the appropriate methods to investigate particular components, a problem of methodology. Scale considerations bear directly on each of these problems. For the latter, a hierarchical perspective that considers an observer's scales explicitly (a hierarchical common sense perhaps more than a hierarchy theory), can guide the choice of appropriate sampling designs and experimental procedures. In addition, hierarchy theory (Allen and Starr 1982, O'Neill et al. 1986) may also provide a useful perspective from which to choose the components of a community that are appropriate for the study of a system's general behavior.

In our discussions here we emphasize three scales that are often used to characterize aquatic systems (Fig. 14.1). Two of these, spatial and temporal scales, are physical and the varied ranges that might be incorporated into an analysis are straightforward. The lengths along the axes represent the extent of space or time considered, while the fineness of rulings on the axes represents the spatial or temporal grain size or resolving power brought to bear in studying a system. Our third axis is conceptually different from, but analogous to the other two and is meant to represent the degree of resolution (or, inversely, aggregation) of entities that investigators might apply to lake features (e.g., criteria for taxonomic or functional groupings). For this third axis we must also consider the extent or proportion of system components that are included in an investigation. With a view towards identifying the best observational scales to use in studying a phenomenon of interest, below we treat each of these axes separately from the point of view of extent and resolving power.

A fourth scale axis pertains to the possible range of approaches that can be used to investigate ecological communities. On one extreme, in-

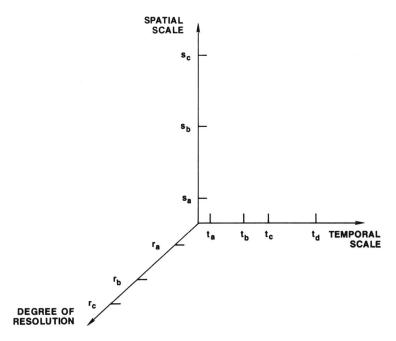

FIGURE 14.1. Three scale axes for consideration in analyses of aquatic communities. A few natural scale lengths are noted on each axis (note logarithmic spatial and temporal axes). For example, t_a, t_b, t_c, and t_d may be diurnal period, annual period, average lifetime of a dominant organism, and mean turnover time of a limiting nutrient in the system, respectively; s_a, s_b, and s_c may be sizes of a small enclosure, embayment, and whole lake, respectively; and r_a, r_b, and r_c may be resolution into trophic levels, functional groups, and species, respectively.

vestigators may observe natural phenomena and infer system behavior from such observations. In contrast, research may involve major manipulations of whole systems. This is a scale of experimental intrusion into a system. Advantages and disadvantages arise systematically along this gradient, also.

In some studies, investigators have broad choices in the scales that they use in studying a system. In others, observations may sometimes be constrained by practical considerations even when a different focus would be appropriate. Typically, scale has not been incorporated explicitly into sampling protocols and experimental designs. Lake study procedures, for example, are frequently dictated by convention, constraints of work schedules, or habit. This has led to the common practice of sampling a lake at one central station to represent parameters occurring in patches ranging in size from μm^3 to m^3 and at weekly intervals to examine processes with turnover times ranging from minutes to months (Harris 1986). In situations where alternate sampling regimes might reasonably be substituted, scale considerations can guide such choices. Alternatively, when sampling protocols reflect practical or methodological limitations, scale consideration can help illustrate limitations in data and avoid errors in interpretation.

In this report, our purpose is to illustrate how scale choice influences our perceptions of aquatic systems and how scale consideration can help guide the unravelling of complex interactions in lake communities. We attempt to provide an overview of the range of scale choices that are involved with aquatic studies, some guidance to investigators in choosing appropriate scales where choice is possible, and suggestions for circumventing scale problems when choice is constrained by the system under investigation. These attempts are in large part practical in trying to provide the best answers to questions that investigators pose about communities. In addition, we hope to provide some insight into the ways that scale considerations may help to guide the choice of questions to be asked by community ecologists.

Temporal Scales

Appropriate time frames for viewing aquatic communities are dictated by a variety of physically imposed periodicities (e.g., diurnal, annual, and longer) and by features of biotic community components. Consideration must be given to the average lifetimes of organisms and the turnover times of energy and nutrients within community components. In addition, observational time scales must allow for features that arise from complex interactions within the community, such as the time for propagation of a signal (e.g., an experimental perturbation) through a web of interactions. Concern must be paid both to the temporal extent, or duration over which a study is conducted, and to the temporal resolution, or sampling frequencies that are used during a study period. In either case, the time scales over which a pattern or process is evaluated can have a major influence on the way that it is interpreted. Some of the consequences of an inappropriate temporal extent, such as a study period that is too short, are obvious.However, the choice, a priori, of an appropriate study length is not obvious. The importance of temporal resolution or sampling frequency may be less conspicuous, but Carpenter and Kitchell (1987) demonstrate it clearly in a model relating primary production and zooplanktivory by fishes. They found that the correlation between primary production and zooplankton biomass changed from a significant negative relationship to a significant positive one as the time scale of sampling was changed from 3 to 7 days (Fig. 14.2). Overall, errors can develop when samples are taken with a frequency that is either too narrow or too broad relative to the response time or periodicity of the phenomenon of interest.

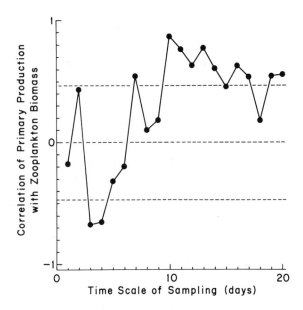

FIGURE 14.2. Pearson product–moment correlation coefficient of primary production with zooplankton biomass vs time scale (days) for a time series derived from a model. Dashed lines: upper and lower denote statistically significant correlations ($P < 0.05$); middle line denotes zero correlation. (From Carpenter and Kitchell 1987. © University of Chicago Press.)

Overly Long Time Scales

Measurements of nutrient availability illustrate the problems that can arise when a sampling period is too long. When phosphorus is released by feeding zooplankton it is immediately taken up by algae (Lehman and Scavia 1982). Only a measure indicating phosphorus concentration during the period between release and uptake would provide a reasonable indication of P availability. Numerous processes occurring in the microbial loop are similarly difficult to evaluate because the shortest time in which a measurement can be made is longer than the time over which a process occurs. Sampling at time scales that are too long can also result in some behaviors, such as oscillations, being missed (Kempf et al. 1984). In some cases, these problems can be rectified by shifting sampling frequencies or by using experimental approaches that alter the net effects of cycling time. However, in other instances processes occur very quickly relative to our abilities to perceive them. Where methodological or practical limitations preclude shifting measurement frequency, simulation models in which the time period for pooling is varied systematically may provide insight into the implications of varied time scales.

Overly Short Time Scales

Difficulties can arise when a study period is too short relative to a phenomenon of interest. Responses to large-scale manipulations typically take
place over several years. Abundance of adult lake trout appeared unaf-
fected by acid additions during the first years of a whole-lake experiment,
but continued studies revealed substantial acid effects during later years
(Schindler et al. 1985). The response of Lake Washington to sewage diversions occurred, at the least, over several years (Edmondson 1972).
The interpretation of smaller-scale experiments will also vary in response
to the time period over which the experiment is carried out. The addition
of a predaceous cyclopoid to tanks containing only herbivorous calanoids
initially depressed the abundance of calanoids through predation. After
several months, however, calanoids were more abundant in tanks with
cyclopoids than in those without as a result of a better persistence of
edible phytoplankton populations in tanks with cyclopoids (D. Soto, pers.
comm.).

Sampling frequency can also involve time scales that are too short. At
the least, sampling too frequently can result in an inefficient use of resources. If other sources of variation, e.g., spatial distribution or taxonomic
resolution, are neglected as a consequence, sampling over short intervals
could be directly linked to misinterpretations. Analysis that preserves too
much fine-grained information can destroy patterns attributable to large-
scale phenomena. Aggregation at the analysis stage is an important scaling
protocol.

It may seem that the problem of a short observation scale is a simple
one that can be addressed by conducting an experiment or making an
observation over a sufficiently long time. More generally, however, if a
basic goal of ecology is to understand the importance of processes that
dictate the patterns observed in natural systems, it is critical to observe
the full range of such patterns and processes occurring in a habitat. This
begs the basic question, "How long is it necessary to observe a system
in order to develop an estimate of the factors that influence it or to predict
its future behavior?". Answering this question requires a choice of an
appropriate time scale for observations, and ideally, several replicate observations of system responses to key periodic events or manipulations.

Choosing Appropriate Time Scales

Factors related to time scale choice include the life spans of the organisms
that comprise a community and the potential time courses of periodic
cycles and stochastic events that occur in a habitat. Diel cycles impart
perhaps the most basic periods characterizing aquatic communities. For
some short-lived organisms, the day-night cycle and associated shifts in
physical conditions may explain most of the variability that occurs over
numerous generations. Longer-lived organisms incorporate numerous day-

night cycles within a lifespan. Observations of certain bacteria over several 24 h periods may describe the full range of their behavior, while replicate observations over the same period would never be used to infer the processes affecting a fish population.

The range of conditions that occur within a year vary with less regularity than a 24 h cycle, but are still predictable. Physical mixing patterns and related chemical features have strong annual signals in temperate (Wetzel 1983) and even tropical lakes (Lewis 1983). Numerous aquatic organisms track seasonal cycles. Lake turnover and subsequent stratification are often followed by a regular sequence of phytoplankton (Reynolds 1984; Sommer 1985) and zooplankton (Threlkeld 1979; Tessier 1986). Such regular sequences can be observed even when turnover and stratification occur without major shifts in temperature, for example, in tropical lakes (Lewis 1978) and experimentally manipulated systems (Reynolds 1984). Several authors (Margalef 1968; Sommer 1985; Harris 1986) have treated turnover in planktonic systems as an analog of characteristic physical events in terrestrial systems, such as fires and blowdowns, which serve to reset biotic interactions in a community (Pickett and White 1985). To the extent that such physical events exert a major organizing effect on processes in communities, they seem an appropriate time scale over which to make repeated observations. This fact and the regularity in seasonal patterns suggest that years or turnover events can provide an appropriate level of replication for examinations of some aquatic community patterns.

Additional important features of aquatic ecosystems are affected by events with longer than annual periodicities. Most fish species, for example, exhibit life histories several years in length. Carpenter (chapter 8) has shown that such life histories may impart variability patterns with multi-year periodicities. Thus, the range of direct interactions between herbivorous zooplankton and phytoplankton during a seasonal cycle may be ascertained over a few years of sampling. However, more complex interactions in which fish influence phytoplankton indirectly through direct effects on zooplankton require a longer time frame. Data derived from periods in which only a restricted range of fish-zooplankton interactions take place may yield interpretations that do not reflect the full range of the processes occurring within a community. Mills (this volume) describes major shifts in perch-*Daphnia* interactions that occur over intervals greater than 5 yr in length. Tonn and Magnuson (1982) demonstrate that winter-kill events with multiyear frequencies can control fish community structure. In discussing classic long-term experiments, Likens et al. (1977) and Schindler et al. (1985) describe how changes occurring unexpectedly after greater than 5 yr of continuous study led them to major shifts in their interpretations of ecosystem function.

The answer to the question of how long to study a system can never be answered simply. An opportunity to observe repeated responses to dominant natural patterns should be a primary consideration in order to place a particular event or process within the perspective of the range of factors influencing a community.

Analyses of Replicated Temporal Cycles

Having chosen an appropriate time scale for observations or experimental manipulations, one must also consider the potential errors associated with assessing one or a limited number of responses to a particular event. Clearly, no investigator would claim an understanding of interactions between a predator and a prey species after an experiment run once in one jar. However, a 5-yr study of the interplay between yellow perch and zooplankton may represent the impact of one strong perch year class and as such parallels a one jar experiment. An evaluation of the relationship between perch and zooplankton would actually call for several, 5-yr observation periods before the range of interactions could be described with confidence. Many experiments and observations in natural systems may require time frames that are similar to this hypothetical example.

Ideally, this problem could be addressed by long-term experiments or observations on several similar systems. Practically, and particularly in the case where long-term data cannot be obtained, the potential range of natural conditions must be inferred to provide a baseline against which to measure a particular process. A combination of simulation models, paleolimnological records, data from other habitats, and an understanding of basic ecological principles can be combined to estimate a natural baseline. Analyses of multiyear data from a set of lakes in the same region may be particularly useful in describing temporal and spatial variability patterns for limnological parameters (Kratz et al. 1987). Some estimate of a long-term baseline is necessary to place the results of any experimental manipulation into a natural context.

Time Scales and the Design of Experiments

The identification of appropriate time scales for examining community processes is particularly important because it affects what can be learned about the community by performing various types of experiments. Questions arise regarding the period necessary for observations and the time frame in which treatments are applied.

Bender et al. (1984) illustrate the importance of temporal considerations in contrasting two designs in experiments involving perturbations; Press and Pulse. Press experiments are those in which a long-term exogenous forcing is applied to each of the accessible components of a community and the resulting steady state values of each of the other components is measured. The external forcings may be an artificial harvesting of a component to reduce its level or the continual input of new members of that component to increase its level. A Pulse experiment involves a short perturbation applied to a component (e.g., an addition or subtraction of some community members) followed by measurement of rates of change in other community components immediately after the pulse.

The choice between a Pulse or Press approach will depend upon the temporal response time of components in the community under investigation. Temporal response time is a scale length that refers to how long it takes a component to react to some perturbation. For example, the doubling of the steady rate of consumption of phytoplankton by a herbivore may ultimately reduce the biomass of phytoplankton by 20%, but it may take 2 wk for the 20% reduction to be approached. The 2 wk is roughly the response time, Tr, of the phytoplankton to that perturbation. A more precise definition of Tr might be the amount of time required for the phytoplankton to decline to within 10% of a steady state under the experimental conditions.

Consider the hypothetical case in which one wants to estimate the strength of interaction between two species, 1 and 2, competing exploitatively for a resource, say the phytoplankton mentioned above. Assume further that species 2 will respond instantly (within days) to a change in the resource level. A Press experiment in the form of doubling the level of species 1 will reduce the resource to 80% of its original value after 2 wk, which will lead to a reduction of the other competitor, species 2, to 80% of its original value over the same period. This experiment plus a corresponding Press applied to species 2, with a measurement of species 1, will allow an assessment of the interaction between the two species (e.g., the interaction coefficients, a_{ij}'s, of competition theory). In contrast, a Pulse reduction of species 1 might last for only a few days and have insufficient effects on the resource base to affect the abundance of species 2. Thus, a Press experiment would be required to show the interplay between these species. If, in contrast to this case, the resource responds very quickly to changes in the level of consumption, say a doubling of consumption by species 1 reduces the resource level to 80% of its original value within a day, then a Pulse experiment would yield measurable results.

The preceding discussion can be formalized by mathematics. The point to be made here is the qualitative one that the intrinsic time scales of the components in an ecological community have implications for the type of experiments that must be performed to determine interactions in the community. It is also important to note that the concepts of Pulse and Press are themselves time scale dependent. A Pulse experiment for some organisms in a lake could be a Press experiment for other organisms with faster turnover times.

Beyond concerns for the design of experiments, the time scales used for experimental manipulations may affect our interpretations of basic ecosystem properties. Consider an idealized aquatic food chain; for example, Figure 14.3. In this chain, it is commonly the case that the responses of the higher trophic levels tend to take longer than those of the lower levels (Stein et al. this volume). For this reason a short-term pulse in the lowest level (typically short time scale phytoplankton species) is rapidly

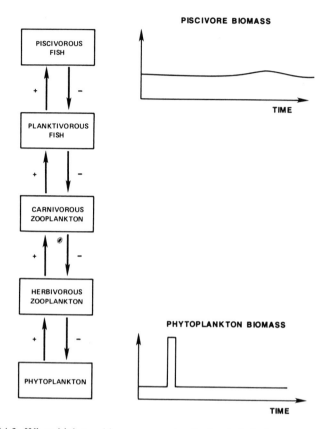

FIGURE 14.3. When high trophic components of a food chain have slower response times than the lower level components, a pulse is unlikely to propagate up the chain.

attenuated as it moves up the food chain, especially when it reaches the piscivorous fish component. Only a long-term perturbation in the primary producer level will show a response at upper levels. On the other hand a relatively short-term perturbation to the piscivorous fish will probably be detectable at the primary producer level. Thus, Pulse experiments may not be able to detect a bottom-up perturbation of food chain dynamics, but top-down effects would be clearly evident. Press experiments carried out over time periods of the order of the lifetime of the piscivorous fish would be needed to detect bottom-up aspects of control.

Similar considerations are necessary in more complex situations, such as looking for indirect interactions between top carnivores in two parallel food chains (Fig. 14.4). It seems likely that a Pulse experiment increasing consumer 1 in the system would have an indirect positive effect on consumer 2. However, this might be impossible to show in the time frame of a Pulse experiment.

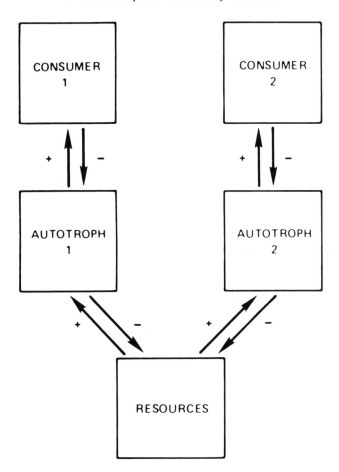

FIGURE 14.4. Parallel trophic chains in an ecological community. A Pulse experiment performed on consumer 1 may not be detected at consumer 2 because different time scale components in the network cause the perturbation to be attenuated.

In addition to a concern for how an experimental manipulation is applied, it is important to consider the timing of such an application in relation to the natural occurrence of a particular pattern or process. Bartell (this volume) has shown that the same perturbation applied to model pelagic food webs can have radically different effects depending upon when it is applied, even at times when parameters describing the system are identical.

Spatial Scales

Spatial scales are imposed by the physical and geological characteristics of a lake; total lake size, depth, depth of the epilimnion, size of basins within a lake, size of littoral zone, large scale gradients of nutrients and oxygen, etc. Other scale lengths are influenced by the biota; organism patch sizes, habitat area sizes, local nutrient concentration patches, and so forth. How do these characteristic spatial scales influence our ability to measure and understand community structure? Answering this question requires a consideration of the extent over which a phenomenon occurs and the sampling methods that are employed to assess it.

Spatial extent and heterogeneity are indispensable for the persistence of a complex ecological community (DeAngelis and Waterhouse 1987). Communities occupying small, spatially homogeneous areas are subject to two basic types of disruption; stochastic environmental disturbances causing random extinctions and overly strong biotic couplings resulting in drastic population oscillations and competitive exclusions. Spatial extent and heterogeneity can decrease the total effect of these disruptions. If the spatial extent of a system is large compared to the spatial scale on which disruptive forces act, then the ecological community will be subjected to such forces in only some regions at a given time. Following a disruption, the space left vacant or depauperate in species can be recolonized from parts of the community in other regions. The conditions under which the spatial area is of sufficient scale to ensure community persistence has been explored, with mathematical models, by den Boer (1968), Reddingius and den Boer (1970), and Roff (1974a, b), among others.

Spatial extent and heterogeneity can also smooth out the population oscillations and extinctions that can occur because of overly strong biotic coupling. These oscillations and extinctions will normally only occur locally and in parts of a lake where factors such as vegetative cover do not moderate the coupling. When spatial extent is great enough, and dispersal rates of prey species or competitively inferior species are great enough, the community as a whole can continually compensate for local dynamic instabilities. This situation has been modeled in a large number of different ways (Levins 1969; Cohen 1970; Levins and Culver 1971; Vandermeer 1987; Slatkin 1974; Hastings 1978; Armstrong 1976; Levin 1976; Yodzis 1978; Allen 1983; Hilborn 1975; Gilpin 1972; Caswell 1978; Zeigler 1977; Hogeweg and Hesper 1981).

Underlying the above considerations is the fact that ecological systems tend to act less as homogeneous wholes than as a mosaic of patches that are often weakly connected on short time scales. A very small patch, or spatial cell, will never be close to equilibrium and, in fact, it is improper to even define an equilibrium state for such a patch. An example from forest community theory may make this idea concrete. Simulation modelers have chosen the 0.1 ha scale as an appropriate one to model tree community dynamics on an individual by individual basis (Botkin et al. 1972; Shugart and West 1977; Shugart 1984). The 0.1 ha patch is a nonequilibrium system; it alternates between periods of successional dynamics (strong biotic coupling causing competitive exclusions) and disturbances (tree-falls) that return the patch to an early state colonizable by a variety of species, including the competitive subdominants. There is no steady state for the patch. The same conceptual and modeling approach has been used for describing the rocky intertidal system (Levin and Paine 1974, 1975; Paine and Levin 1981), where approximately 1 m^2 is used as the patch size. In these approaches the forest and rocky intertidal communities as a whole are conceived to be made up of a large number of nonequilibrium patches, coupled through the dispersal of seeds or propagules. This coupling is extremely important over the long time scale but over short time scales the patches may be relatively independent.

A lake can be conceived of in a way that is fundamentally very similar to the preceding conceptualizations of the forest and rocky intertidal systems, although the relevant spatial and temporal scales of the patches might be quite different. One major consideration involves fundamental differences between lake patch dimensions on vertical and horizontal scales. Vertical patches (strata) range in size from the depth of the epilimnion to layers a few centimeters thick in the hypolimnion, while horizontal patches may occur much more broadly (Harris 1986). Patch size and persistence will also vary with the parameter that is being used to define it.

An understanding of patch size and dynamics could be helpful in several general ways. On a basic level, it can help in the understanding of the structure of ecological communities and the processes that maintain it. On a practical level, spatial scale considerations will affect the way that lake patterns can be inferred from samples. Also, explicit considerations of spatial scale are necessary before experiments done at small spatial scales can be related to the properties of a complex community in a whole lake.

Sampling and Lake Patterns

As was the case for temporal scales, it is possible to employ a sampling protocol that is either too broad or too narrow for the parameter under consideration. Sampling devices in a broad variety of sizes have been used in aquatic research and, we suspect, sampler size is often dictated by what is available or convenient rather than the pattern or process under investigation.

In some cases the implications of sampling too large a scale are obvious. For example, a complex array of 1 cm thick layers of 10 hypolimnetic phytoplankton species (e.g., Watras and Baker in preparation) would be reduced to mixed assembly of all species when a sample was taken using a 1 m long VanDorn Bottle. Similarly, a net towed through an epilimnion to sample zooplankton will provide no information on vertical distribution patterns.

On a more fundamental level, Folt (1987) has described how the use of a sampler that is not scaled appropriately to the aggregations of study organisms may lead to important differences between the apparent density of an organism and the density that it actually experiences. For example, zooplankton samples taken with a 30 l Schindler-Patalas Trap may suggest densities for *Mysis* and *Daphnia* of 20 individuals per/l for each taxon. If, in actuality, the mysids occurred in a thin band at the bottom of the chamber and the daphnids in a similar band (but at the top of the chamber), the densities experienced by the organisms would be underestimated while the degree of interaction between the two populations would be overestimated.

Patchiness also affects samples taken at too narrow a range compared to a pattern or process of interest. Peristaltic pumps may sample from thin layers and miss other strata in which important processes are taking place. Abundant zooplankton in patches may be missed when samples are taken at a central lake station (Tessier 1983) and horizontal heterogeneity is ignored.

In some cases, explicit choice of sampler size and overall survey design can circumvent problems with sampling schemes that are too broad or too narrow. Where appropriate approaches are not predictable or practical, it may be useful to employ simulation models of lakes that incorporate varied spatial aggregation scales to indicate the consequences of sampler design.

Small Scale Experiments and Whole Lakes

Experiments performed in enclosures that are small relative to lakes have provided numerous critical insights into aquatic community processes. Results from enclosures, however, must eventually be related to processes occurring at the scale of a whole lake if their significance to natural communities is to be understood. Physical size is one major factor affecting the relationship between enclosure results and natural community patterns. How small an enclosure actually is, relative to a whole lake, depends, of course, upon the organisms or processes under consideration. However, after a period of time, T, an enclosure of spatial scale L will depart drastically in community structure from the lake as a whole due either to stochastic extinctions or to overly strong biotic coupling leading to extinction. For many processes, it is likely that the divergence time, T, will be proportional to L.

The divergence with time of a small enclosure from whole-lake conditions, is, in part, a result of the loss of the compensatory effect of spatial extent and heterogeneity previously discussed. Extrapolations from the results of small enclosure experiments to the community dynamics of whole lakes may be validated by using mathematical models to factor in the possibility of periodic reinvasion of a patch or cell from nearby cells and of dispersions from the cell of populations that have become too dense. Such an extrapolation would require that the dynamics in an enclosure be modeled somewhat like the forest patch or cell models of Botkin et al. (1984) and Shugart and West (1977), and that the coupling of cells also be properly modeled based on a knowledge of dispersion of various types of organisms in lakes.

By means of a graded series of enclosure sizes, it may be possible to derive a relationship between enclosure size and the time scale over which the enclosure exhibits processes that parallel those in a whole lake. Such information could allow the derivation of scaling laws to directly relate data from experimentally manipulable enclosures to the whole lake community.

Scale of Resolution of Components

The degree to which organisms or processes associated with organisms are resolved in the evaluation of an ecological community (or its opposite, the degree to which community components are aggregated) is viewed here as a question of scaling. Just as studies can be carried out over varied ranges of spatial scales and temporal scales, so too can the resolution of the components used in the description of the ecological community be varied. In addition, having decided upon scales for resolution, the proportion of the total components in a community that are included in an investigation must also be considered directly.

Components is a general term that can refer to taxa, size classes, functional groups, or any of a large number of relevant categories (Fig. 14.1). It should be emphasized that the determination of components in a system is an arbitrary decision of the observer. For example, in trying to test theories for diversity in ecological communities, a high degree of taxonomic resolution is essential. On the other hand, when examining bottlenecks and other features of energy flow up the food web, size class resolution may be critical (cf. Neill this volume).

It is clear that high resolution of system components is often a major goal in the investigation of ecological communities. However, it is not always possible, and may not actually be necessary in all situations. Furthermore, there may be some cases where too fine a degree of resolution may mask the detection of community-level processes. There are several reasons for aggregating the components of a community including:

1. The phenomenon of interest may imply a high degree of aggregation. For example, in studies of primary productivity the contributions of many autotrophic taxa are lumped.
2. Constraints on the practicality of sampling and measurement may make a high degree of aggregation necessary. For example, approximate total numbers of fish may be obtainable by sonar devices, but species identification might not be possible.

3. Aggregation may be useful in detecting regularities and inferring general laws about ecological systems. For example, the total biomass in a guild or functional group of species in a community may remain relatively constant or change predictably, whereas individual species in that group may fluctuate erratically (Tonn 1985). Moreover, comparisons can be made for aggregations, such as trophic groups, between ecosystems that share no species in common (Stein et al., chapter 11).

4. For most natural communities, considerable aggregation is necessary before any mathematical modeling of the system's dynamics can be attempted. The community must be reduced to a small number system.

Over-Aggregation or Exclusion of Community Components

When community components are lumped arbitrarily, or observations do not allow fine enough discrimination between components, the resulting aggregation can sometimes lead to errors. Aggregations that obscure major system components can lead to a basic misunderstanding of community behavior. This possibility was clearly demonstrated in Neill's paper (Chapter 3) where the lumping of rotifers with other herbivorous zooplankton in a lake's food web would have led to an erroneous interpretation of the factors controlling *Chaoborus* population dynamics.

The question of how many system components (e.g., species) must be included in evaluating a community is also related to aggregation. In some cases, aggregation can involve the inadvertent lumping of system components into broader taxonomic or functional groups. Alternatively, only a subset of system components is considered with others excluded either by accident or by design. Rare species may simply be missed, or an investigator may choose to examine specific groups such as producers or predators. In either situation, where the line is drawn between recognized and unrecognized groups can have a critical effect on community evaluations. Rahel et al. (1984) showed that the assessment by Grossman et al. (1982) as to whether a fish assemblage was affected by stochastic or deterministic processes depended upon the number of species involved in the evaluation. The recent recognition of the microbial loop in aquatic ecosystems is an especially dramatic case of a previously unknown component which may force a reinterpretation of aquatic community dynamics (Stockner and Porter this volume).

A variety of errors can arise through improper aggregation or neglect of components. Estimations of dynamical properties of communities will vary with the degree to which community components are aggregated. Similarly, evaluations of the strength of interactions between sets of components will vary as other components are deliberately or inadvertently ignored. As a first example, consider two species playing similar functional roles in a community (Fig. 14.5). These two species may be lumped into

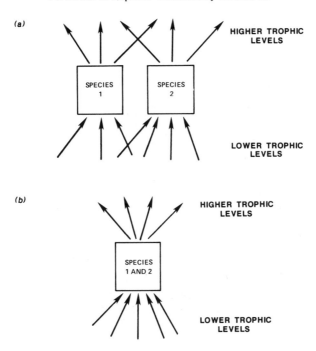

FIGURE 14.5. Hypothetical case of aggregation of two species performing similar functional roles *(a)* into a single component *(b)*.

a single component. However, this aggregation could significantly change the dynamic characteristics of the system measured under some conditions, depending on the relative biomasses of species 1 and 2, their strengths of interaction with other components of the community, and the mathematical details of how the lumping is done. There is a body of literature on the subject of aggregation that may help determine whether or not a proposed lumping of components introduces significant errors (O'Neill and Rust 1979; Gardner et al., 1982; Ziegler 1976, 1979).

Aggregation and the proportion of recognized system components can also influence the inferences that are drawn from manipulative experiments. Moreover, the influence of these factors will vary with the type of experiment involved in a study. Bender et al. (1984, see discussion above) showed that, in experiments designed to measure interactions among community components (Fig. 14.6), the omission of some components can lead to serious errors in Press experiments, but that such errors are less likely in Pulse experiments. In this case, Pulse experiments, in contrast with Press experiments, are insensitive to long-response-time effects from the overlooked or ignored components. Thus, Pulse exper-

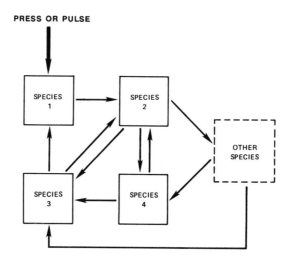

FIGURE 14.6. Hypothetical situation in which an experimental manipulation (Press or Pulse) is proposed to elicit the interaction strengths among species 1, 2, 3, and 4. The existence of species not explicitly considered ("other species") leads to errors.

iments will be useful in measuring direct effects in experimental manipulations even when some community components are not recognized. However, they will be of limited utility in identifying indirect or complex interactions in aquatic communities because they are not responsive to processes with long lag times. In addition to problems with missing system components, the degree to which species are aggregated will also influence the utility of pulse, press, and other experimental designs.

Analyses of food web structure are highly susceptible to problems of overaggregation. Frequently, such analyses (Briand 1983) involve direct comparisons of food web components aggregated at levels that vary from individual species to broad functional groups. Assessments of interactions, such as connectance, drawn from such varied aggregations may be a property of the groupings used rather than of the community under investigation. In some cases, studies of the complexity of food webs may be able to more realistically evaluate the taxa involved at all trophic levels by incorporating investigators with a breadth of taxonomic expertise. Even where aggregation is unavoidable, however, food web analyses should incorporate specific error analyses that evaluate the consequences of overaggregation for assessments of community parameters.

Under-Aggregation of Community Components

The consequences of measuring community components with too fine a scale of resolution are less obvious than problems associated with overaggregation. On a practical level, efforts expended on fine scale resolution of community components might be better focused on more frequent sampling or larger numbers of sampling stations if the question of interest does not require a detailed knowledge of the fine scale components. It is not necessary to measure the production of individual phytoplankton species in assessing overall system primary production. In contrast, measures of system production will provide little insight in phytoplankton population dynamics.

On a more basic level, variance associated with finely resolved community components may mask patterns in parameters that aggregate appropriate subgroups of components. Reynolds (1984) describes a spring peak of three diatom species in Lake Windermere that is highly consistent when considering the total biomass of the three species. However, the behavior of the three species individually is substantially less predictable than the pattern for the aggregate group. Similarly, Sullivan and Carpenter (1982) have shown that a genus-level index provides a more reliable measure of trophic status than a species-level index derived from the same basic data.

Choosing Appropriate Aggregations

The appropriate degree of aggregation for evaluating an aquatic community depends primarily upon the phenomena under study. The sensitivity of a parameter to change within a community is likely to decrease with the number of species that are aggregated within it. For example, in documenting a lake's response to continuous acid additions over several years, Schindler et al. (1985) observed that primary production showed little change relative to natural fluctuations of production exhibited in similar but unmanipulated lakes in the region. In contrast, certain phytoplankton species disappeared from the lake in the early stages of acidification. Intermediate response times were apparent for parameters that aggregated several species (e.g., the total biomass of a major taxonomic or functional group). While sensitivity is higher in parameters that aggregate fewer species, annual and stochastic variation are also likely to be higher in parameters with a low degree of aggregation. Thus the choice of a parameter or group of parameters to evaluate complex interactions in aquatic communities may involve a compromise between sensitivity and natural variability. This suggests that aggregations of a few species, functional groups, or multivariate measures of community conditions (Caswell and Weinberg 1986; Spiess et al. 1987) may be useful in examining system responses against a background of natural variability.

The aggregation or exclusion of system components can have an important influence on the conclusions that are drawn from observations of a system or from experimental manipulations. Explicit considerations of aggregation scale can help both in choosing the appropriate level of resolution for system components and in selecting methods of experimentation and measurement appropriate to the level of resolution that one selects.

Scale in the Choice of Research Approach

Important inferences about community structure have been gained from observational studies, experiments in enclosures ranging in size from test tubes to large mesoscosms, and manipulations of entire lakes. There are specific advantages and limitations to each of these approaches. Observational studies have the advantage of examining conditions that are clearly within the normal range for the system under investigation. Conversely, they suffer from a limited ability to resolve specific processes against the background of complex and variable natural patterns. Experimental studies often generate strong effects relative to natural variability. However, responses elicited in experiments may have little relation to processes occurring in natural systems. Trade-offs exist, therefore, between generating results that can be discriminated from natural variability and conducting experiments with results that are representative of the natural range of conditions in a study system.

Observational studies, even those conducted over extensive periods, often are not in and of themselves capable of identifying critical community processes. Experimental manipulations are necessary to test hypotheses and to elucidate controlling mechanisms. However, any interpretation of aquatic community processes based upon experimental manipulations must be compared with extensive, natural baseline information to determine its validity in the context of natural fluctuations. Only rarely can these problems be approached with long-term data from several, similar systems. Alternatively, and particularly in the case where long-term data cannot be obtained, the potential range of natural conditions must be inferred. A combination of simulation models, paleolimnological records, data from other habitats, and an understanding of basic ecological principles can be combined to estimate a natural baseline. Some such baseline is necessary to place the results of any experimental manipulation into a natural context.

Experimental Manipulations

The crucial role of experiments in community ecology is clear but there are broad ranges of potential manipulations for each of the scale axes that we have discussed (Fig. 14.1). The choice of experimental design must be driven by the system under study and the questions that are being asked about it. The choice between field or laboratory experiments depends upon the size and life spans of the organisms involved in an investigation and the need to control environmental conditions. Choices for temporal scales vary in several ways including the time at which a manipulation is applied, the period over which the perturbation is maintained, and the time span over which the response to an experiment is observed. Choices for spatial scales range from test tubes to whole lakes. Choices for aggregation and/or exclusion of system components in experiments concern the portions of a system that are manipulated and the portions that are used to measure a response. Manipulations can range from the reduction of a single species to the removal of major portions of a trophic level. Similarly, responses can be assessed either in terms of single species or processes such as primary production. Any experiment should include an explicit consideration of the scales that it involves and the consequences of scale choice for the interpretation of its results.

Relative to a system's natural variability, an investigator's ability to detect responses to a perturbation will increase as the impact of the manipulation increases. At the same time, however, information on responses to relatively minor perturbations may be desirable. This is particularly the case in studies that attempt to detect the early effects of environmental contaminants. An investigator must contrast sensitivity and natural variability in choosing the magnitude of a manipulation employed in an experiment.

Whole-Lake Experiments

Whole-lake manipulations have major advantages in the ways in which they simulate, or actually encompass, the conditions that would be expected to occur naturally in a lake. Problems of enclosure size or missing members of communities are not relevant in whole-lake manipulations. However, large-scale experiments are also plagued by problems that are specific to their scale. An obvious shortcoming is that, because of logistic considerations, the scale of replication in such experiments is typically limited to one lake or two half-lakes (Hurlbert 1984). At the same time, whole-lake experiments usually are run over extensive periods under natural conditions and as such have a large potential for temporal and between-basin variability. Thus, treatment effects cannot be assessed by straightforward statistical analyses. The fundamental question in any experimental manipulation, "What would have happened in the treated system if the treatment had not been applied?", is very difficult to answer for whole-lake studies.

Although experimental design has been a problem in some work (Hurlbert 1984), smaller-scale experiments using enclosures or tanks can be conducted with replication. Despite replication and the use of standard statistical approaches, however, the relationship between enclosure experiment and natural patterns and processes is not always clear. Considering the potential advantages of whole-lake experiments for the interpretation of complex community processes and the difficulties in interpreting their results, an exploration of methods for evaluating whole-lake manipulations is appropriate here.

The basic design of most whole-lake manipulations involves repeated measures of condition in a treatment system and in at least one reference system. In most studies the relationship between the treatment system and reference system will be estimated in a baseline period prior to experimental manipulation. Within this framework, the questions asked in whole lake experiments can be broken down into three levels.

1. In a comparison between a reference and a treatment system in a specific experiment, is the magnitude of the difference(s) between the reference lake and the manipulated lake during the post-manipulation period significantly different from what it was during the pre-manipulation period?
2. If so, can the change be attributed to the manipulation rather than to other factors affecting the basins?
3. Finally, how do results from the particular systems under study relate to lakes in general?

In testing for differences between the specific reference and treatment basins in an experiment (question 1), typical measures involve repeated comparisons over time during both a baseline and post-treatment period. One major difficulty in such comparisons is the potential lack of independence in samples collected repeatedly from the two basins. Weekly estimates of fish populations in a lake basin would normally be strongly correlated week by week. On the other hand estimates of young of the year fish populations made at yearly intervals are often affected by essentially independent processes and show little correlation year by year. A basic approach to obtaining independent data would involve the use of a sampling frequency that was longer than the predominant cycling times of the phenomenon being measured. A variety of statistical procedures, many of which have not typically been employed by ecologists, may be appropriate for comparisons of reference and treatment basins over time. These include randomization tests (Edgington 1980), Time Series Analyses (Box and Jenkins 1976; Hirsch et al. 1982; Hirsch and Slack 1984; Oehlert 1984) and procedures to control for non-independence in data (Burgess and Webster 1980a, b; Robertson 1987). Statistical techniques developed for epidemiological studies, which involve analytical problems with numerous parallels to those in whole-lake experiments, may be particularly useful (Lilienfeld and Lilienfeld 1980; Schlesselman 1982). Other potential approaches that are specifically intended to explore whole-system responses include a procedure suggested by Stewart-Oaten et al. (1986) for environmental impact assessments and a procedure combining intervention analysis (Box and Tiao 1976) and randomization suggested by Carpenter et al. (in preparation) for whole-lake experiments.

Regardless of how clear a difference appears between a reference and a treatment system after a manipulation, lack of replication makes it difficult to unequivocally attribute such a difference to the experimental treatment (question 2). For example, consider a lake that has been divided in half by a curtain and acidified 0.5 pH units after a 1-yr baseline period. If dramatic differences in zooplankton populations develop between the two basins after acid additions, it may seem reasonable to link such differences to the effects of added acid. However, if by chance, a strong year class of a zooplanktivorous fish developed in the acidified basin but not in the reference basin, substantial changes might occur in zooplankton that are unrelated to pH. Moreover, a strong year class of fish may affect zooplankton for several years, perhaps for a period longer than the planned course of an experiment.

Other random events in the two separate basins of a divided lake could also confound the interpretation of a whole lake manipulation. In addition, unidentified factors (e.g., differential groundwater inflow or outflow) may have unequal effects on halves of a divided lake. Such confounding factors can be a particular problem when an experiment is designed to examine perturbations that are not dramatic relative to natural lake conditions. Essentially, researchers must ask if they are likely to obtain the same result in a whole-lake experiment if they were to set everything back up and run the experiment over again. Alternative explanations for observed differences in whole-lake manipulations must be given explicit considerations.

One direct approach to the sample size problem in whole-lake experiments could involve increasing the degree of replication. Logistic, practical, and aesthetic considerations will limit this option in many cases. However, the use of several reference systems to contrast with a single treated system may circumvent some of these difficulties. Schindler et al. (1985) employed such an approach in assessing the results of a long-term lake acidification experiment. The utility of a set of reference lakes ultimately depends upon the nature of the variability exhibited by a parameter of interest.

Considering long-term data for a set of lakes within the same region, variability patterns for parameters measured simultaneously for each lake can be classed into three categories. Parameters may be lake-specific, consistent in each lake over a number of years but variable among lakes; year-specific, consistent across lakes in any year but variable among years; or complex, consistent neither across lakes or years (Kratz et al. 1987; Frost and Kratz in preparation). A parameter may be both lake- and year-specific. Predictions of the conditions in a manipulated lake basin, based on a set of reference lakes, are most likely to be reliable for parameters that show year-specific patterns of variability. At the same time, changes in lake-specific parameters may also be attributable to a manipulation, since they are unlikely to change in a lake under natural conditions (Kratz et al. 1987). Interpretations of complex parameters will be difficult even if data are available from a large set of reference lakes. Thus, the analysis of variability patterns in a set of lakes established as a reference for an experimental manipulation or, more generally, in a set of lakes under study within a region can provide important guides to the interpretation of whole-lake experiments.

An alternative experimental design for whole-lake experiments involves sequential treatments of two basins with a substantial lag time between the first and second treatment (Brezonik et al. 1986). Initially, this approach follows the typical design of a whole-lake experiment in which comparisons are made between a reference and a treatment basin. However, the approach diverges from traditional methods when the experimental manipulation is repeated in the reference basin one or more years after the initial treatment. The power of this approach is in its ability to confirm that results observed in response to the initial treatment are repeatable in a second system. If the same result is observed in both treatment 1 and treatment 2, despite interannual and random differences between the basins when the treatments were applied, it can be attributed with some confidence to the experimental manipulation. Alternatively, if a result is not repeatable it will not be attributed unjustifiably to the experimental treatment, but will suggest areas for further study. The staggered replication design illustrates the need to employ nontraditional approaches in assessing whole-lake experiments.

The relationship between the response in a particular whole-lake experiment to aquatic communities in general (question 3) must be based upon an identification of the mechanisms that lead to the patterns observed in a particular lake. It cannot be assumed that the lake in which an experiment is conducted is a random representation of lakes in general. The staggered replication design and the use of a set of reference lakes discussed above will also provide insight into the generality of the response in a particular experiment.

Ultimately, the value of whole-lake manipulations may be highest when they are used in a hierarchical set of experiment carried out on a variety of scales. Patterns that appear in contrasting treatment and reference systems in whole-lake experiments can be used to suggest important system processes. Subsequently, the importance of an inferred process can be tested in smaller-scale experiments that can be replicated and tested statistically. In this approach, whole-system experiments can be considered as generating as well as testing hypotheses.

The utility of whole-lake experiments is clear. Their use in conjunction with smaller-scale experiments can provide insights into community processes that would not be available from studies conducted solely at one scale or the other. The practical difficulties of conducting numerous whole-system manipulations suggests the value of cooperative programs involving several investigators taking advantage of each whole-system experiment.

The success of the programs at the Experimental Lake Area in Canada and at the Hubbard Brook Experimental Forest in New Hampshire (Likens 1985) illustrates the value of such an approach. Such cooperative studies can be further enhanced if they are conducted in areas where substantial reference data on unmanipulated systems are also available. Finally, it will be useful to view the numerous manipulations that are frequently performed in a variety of management practices (e.g., fish additions and removals, nutrient diversions, reservoir constructions) as opportunities for whole-lake experiments with minimal logistic costs. In addition, natural system perturbations (e.g., winterkill-Tonn and Magnuson 1982) may provide opportunities for natural experiments (Diamond 1986) on whole systems.

Conclusions

Like other sciences, community ecology searches for regularities in nature, tries to understand these regularities, and attempts to use this understanding to predict future behavior. Every study of an ecological community is, in truth, the study of a highly abstracted and simplified version of that community. For practical reasons, we can perturb and sample the community only over limited ranges of space and time and only with a limited degree of resolution for the entities composing it. The task, then, is to make our studies of communities as efficient as possible by intelligent choices of scales on which to conduct them. To do so it is necessary to identify the characteristic scales at which regularities are likely to occur and ultimately, the dominant scales operating within a community. We hope this chapter has provided a few useful guidelines with respect to scale in aquatic communities.

Summary

The temporal and spatial scales that exist for limnetic communities result from basic physical and geological conditions (e.g., diurnal and annual cycles, like morphometry) and biotic dynamics (e.g., organism life spans, nutrient turnover times). For aquatic communities, these factors interact to create time scales ranging from fractions of a second to decades, and spatial scales from fractions of centimeters to tens of meters. In addition, the degree to which entities in a community are perceived or aggregated (e.g., individual species, size classes, functional groups) can also be considered as a scale perspective. In each case, the scales used to view aquatic ecosystems can have a major influence on the ways that communities are perceived.

Sampling schemes for both observational and experimental studies must be designed with these characteristic scales in mind. It is possible to use a scale that is either too broad or too narrow for a phenomenon of interest. Investigations of the microbial loop, for example, may require sampling schemes on the scale of seconds, while phenomena associated with strong piscivore year classes may require sampling over decades. On spatial scales, analyses of horizontal distributions may be detected using sampling intervals greater than 1 m while vertical distributions require sampling at intervals of a few cm. In resolving community components, aggregation may be a necessity forced by the limited ability of a researcher to collect and measure things, but it may also be an advantage that can lead to the detection of greater regularity in system behavior.

Scale plays an important role in the design and interpretation of experiments. For example, the most effective perturbation for a manipulation will depend upon the response time of the system components under study. Spatial scale presents special problems for experiments, which can usually only be controlled and replicated on scales much smaller than a typical lake. Since it can be demonstrated that any small-scale experimental model of a lake must rather quickly deviate from behavior of the actual lake (at least for complex community processes), methods of extrapolating from behavior on the small scale to that on the larger scale are needed. We discussed a theoretical approach to such a scaling rule. Aggregation must be considered both in terms of the community components that are manipulated in an experiment and in terms of the response variables that are measured to assess a system's response. In some cases, choices can be made for appropriate scale in observations or experiments. In other situations, scale choice may be constrained by practical limitations. Under either circumstance, scale considerations should be incorporated explicitly into any investigation of an aquatic community.

Finally, we considered the scales of manipulation that are employed in community studies, ranging from observations of natural systems through a variety of small-scale experiments to whole-lake experiments. In choosing an approach that is appropriate for a particular question, we discussed how the closeness with which an experimental system approximates natural conditions is balanced against an ability to detect a response against natural variability. Small-scale experiments have advantages in their logistics and replicability, but must be interpreted relative to their ability to reflect natural community processes. Whole-lake experiments incorporate a full range of natural conditions, but are difficult to perform and to replicate. Whole-lake experiments were considered as particularly important, and we discussed techniques for their interpretation. A hierarchical combination of approaches with smaller-scale experiments conducted in the context of whole-system manipulations would appear to provide the most powerful insights into aquatic community processes.

Acknowledgements. We thank T. F. H. Allen for his extensive contributions in the discussions that led to this paper and to the manuscript itself. We also thank Steve Carpenter, Jim Kitchell, Tim Kratz, and Paul Rasmussen for helping to formulate the ideas presented here. T. M. Frost's contributions to this manuscript were supported in part by the North Temperate Lakes Long Term Ecological Research Project (NSF Grant BSR-8514330) and by the Little Rock Lake Experimental Acidification Project (US-EPA Cooperative Agreement #CR-812216-01-0).

References

Allen, L. J. S. 1983. Persistence and extinction in Lotka-Volterra reaction-diffusion equations. Math. Biosci. 65: 1–12.

Allen, T. F. H. and T. B. Starr. 1982. Hierarchy: Perspectives for Ecological Complexity. Chicago: University of Chicago Press.

Armstrong, R. A. 1976. Fugitive species: experiments with fungi and some theoretical considerations. Ecology 57: 953–963.

Bartell, S. M., T. F. H. Allen and J. F. Koonce. 1978. An assessment of principal component analysis for description of phytoplankton periodicity in Lake Wingra. Phycologia 17: 1–11.

Bender, E. A., T. J. Case and M. E. Gilpin. 1984. Perturbation experiments in community ecology: Theory and practice. Ecology 65:1–13.

Botkin, D. B., J. F. Janek and J. R. Wallis. 1972. Some ecological consequences of a computer model of forest growth. J. Ecol. 60:849–873.

Box, G. E. P. and G. C. Tiao. 1975. Intervention analysis with applications to economic and environmental problems. J. Amer. Stat. Assoc. (Theory and Methods Sec.) 70:70–79.

Box, G. E. P. and G. M. Jenkins. 1976. Time Series Analysis: Forecasting and Control. San Francisco: Holden-Day.

Brezonik, P. L., L. A. Baker, N. Detenbeck, J. Eaton, P. Garrison, T. Kratz, J. Magnuson, J. Perry, W. Rose, B. Shepard, W. Swenson, C. Watras and K. Webster. 1986. Experimental acidification of Little Rock Lake Wisconsin. Water, Air, and Soil Pollution 31:115–121.

Briand, F. 1983. Environmental control of food web structure. Ecology 64:253–263.

Burgess, T. M. and R. Webster. 1980. Optimal interpolation and isarithmic mapping of soil properties. I. The semi-variogram and punctual kriging. Journal of Soil Science 31:315–331.

Burgess, T. M. and R. Webster. 1980. Optimal interpolation and isarithmic mapping of soil properties. II. Block kriging. Journal of Soil Science 31:333–341.

Carpenter, S.R. and J.F. Kitchell. 1987. Plankton community structure and limnetic primary production. Am. Nat. 124: 159–172.

Caswell, H. 1978. Predator-mediated coexistence: a non-equilibrium model. Am. Nat. 112:127–154.

Caswell, H. and J. Weinberg. 1986. Sample size and sensitivity in the detection of community impact. IEEE Oceans '86 Conference Proceedings.

Cohen, J. E. 1970. A Markov contingency table model for replicated Lotka-Volterra systems near equilibrium. Am. Nat. 104:547–559.

DeAngelis, D. L. and J. C. Waterhouse. 1987. Equilibrium and nonequilibrium concepts in ecological models. Ecol. Monogr. 57:1–21.

den Boer, P. J. 1968. Spreading the risk and stabilization of animal numbers. Acta Biotheoretica (Leiden) 18:165–194.

Diamond, J. 1986. Overview: Laboratory experiments, field experiments, and natural experiments. *in*: Community ecology, ed. J. Diamond and T. J. Case, 3–22. New York: Harper and Row.

Edgington, E.S. 1980. Randomization tests. New York: Marcel Dekker.

Edmondson, W. T. 1972. Nutrients and phytoplankton in Lake Washington. *in*: Nutrients and eutrophication: The limiting nutrient controversy, ed. G. E. Likens, 172–193. Special Symposium, Amer. Soc. Limnol. Oceanogr. 1.

Folt, C. L. 1987. An experimental analysis of costs and benefits of zooplankton aggregation. *in*: Predation: Direct and indirect impacts on aquatic communities, ed. W. C. Kerfoot and A. Sih, 300–314. Hanover: University Press of New England.

Gardner, R. H., W. G. Cale and R. V. O'Neill. 1982. Robust analysis of aggregative error. Ecology 63:1771–1779.

Gilpin, M.E. 1972. Group selection in predator-prey communities. Princeton: Princeton University Press.

Grossman, G. D., P. B. Moyle and J. O. Whitaker, Jr. 1982. Stochasticity in structural and functional characteristics of an Indiana stream fish assemblage: a test of community theory. Am. Nat. 120:423–454.

Harris, G. P. 1986. Phytoplankton ecology: Structure, function and fluctuation. London: Chapman and Hall.

Hastings, A. 1978. Spatial heterogeneity and the stability of predator-prey systems: predator-mediated coexistence. Theoret. Pop. Biol. 14:380–395.

Hilborn, R. 1975. The effect of spatial heterogeneity on the persistence of predator-prey interactions. Theoret. Pop. Biol. 8:346–355.

Hirsch, R. M., J. R. Slack and R. A. Smith. 1982. Technique of trend analysis for monthly water quality date. Water Resources Res. 18:107–121.

Hirsch, R. M. and J. R. Slack. 1984. A nonparametric trend test for seasonal data with serial dependence. Water Resources Res. 20:727–732.

Hogeweg, P. and B. Hesper. 1981. Two predators and one prey in a patchy environment: an application of MICMAC modelling. J. Theor. Biol. 93:411–432.

Hurlbert, S. H. 1984. Pseudoreplication and the design of ecological field experiments. Ecol. Monogr. 54:187–211.

Kempf, J., L. Duckstein and J. Casti. 1984. Relaxation oscillations and other non-Michaelian behavior in a slow-fast phytoplankton growth model. Ecol. Modell. 23:67–90.

Kratz, T. K., T. M. Frost and J. J. Magnuson. 1987. Inferences from spatial and temporal variability in ecosystems: Analyses from long-term zooplankton data from a set of lakes. Am. Nat. 129:830–846.

Lehman, J. T. and D. Scavia. 1982. Microscale patchiness of nutrients in plankton communities. Science 216:729–730.

Levin, S. A. 1976. Population dynamic models in heterogeneous environments. Ann. Rev. Ecol. Syst. 7:287–310.

Levin, S. A. and R. T. Paine. 1974. Disturbance, patch formation, and community structure. Proc. Nat. Acad. Sci. (USA) 71:2744–2747.

Levin, S. A. and R. T. Paine. 1975. The role of disturbance in models of community structure. *in*: Ecosystem analysis and prediction, ed. S. A. Levin, 56–57. Philadelphia: Society for Industrial and Applied Mathematics.

Levins, R. 1969. Some demographic consequences of environmental heterogeneity for biological control. Bull. Entom. Soc. Amer. 15:237–240.

Levins, R. and D. Culver. 1971. Regional coexistence of species and competition between rare species. Proc. Nat. Acad. Sci. (USA) 68:1246–1248.

Lewis, W. M., Jr. 1983. Temperature, heat, and mixing in Lake Valencia, Venezuela. Limnol. Oceanogr. 28:273–286.

Lewis, W. M., Jr. 1978. A compositional, phytogeographical, and elementary structural analysis of the phytoplankton of a tropical lake: Lake Lanao, Philippines. J. Ecol. 66: 213–226.

Likens, G. E. 1985. An experimental approach for the study of ecosystems. J. Ecol. 73:381–396.

Likens, G. E., F. H. Bormann, R. S. Pierce, J. S. Eaton and N. M. Johnson. 1977. Biogeochemistry of a forested ecosystem. New York: Springer-Verlag.

Lilienfeld, A. M. and D. E. Lilienfeld. 1980. Foundations of epidemiology. New York: Oxford University Press.

Makarewicz, J. C. and G. E. Likens. 1975. Niche analysis of a zooplankton community. Science 190:1000–1003.

Margalef, R. 1968. Perspectives in ecological theory. Chicago: University of Chicago Press.

Oehlert, G. W. 1984. A statistical analysis of the trends in the Hubbard Brook bulk precipitation chemical data base. Technical Report #260 (Series II). Department of Statistics, Princeton University.

O'Neill, R. V., D. L. DeAngelis, J. B. Waide and T. F. H. Allen. 1986. A hierarchical concept of ecosystems. Princeton: Princeton University Press.

O'Neill, R. V. and B. Rust. 1979. Aggregation error in ecological models. Ecol. Modell. 7:91–105.

Paine, R. T. and S. A. Levin. 1981. Intertidal landscapes: disturbances and the dynamics of pattern. Ecol. Monogr. 51:145–178.

Pimm, S. L. 1982. Food Webs. London: Chapman and Hall.

Pickett, S. T. A. and P. S. White, eds. 1985. Ecology of natural disturbance and patch dynamics. New York: Academic Press.

Rahel, F. J., J. D. Lyons and P. A. Cochran. 1984. Stochastic or deterministic regulation of assemblage structure? It may depend on how the assemblage is defined. Am. Nat. 114: 583–589.

Reddingius, J. and P. J. den Boer. 1970. Simulation experiments illustrating stabilization of animal numbers by spreading of risk. Oecologia (Berlin) 15:259–275.

Reynolds, C. S. 1984. The ecology of freshwater phytoplankton. Cambridge: Cambridge University Press.

Robertson, G. P. 1987. Geostatistics in ecology: interpolating with known variance. Ecology 68:744–749.

Roff, D. 1974a. Spatial heterogeneity and the persistence of populations. Oecologia (Berlin) 15:259–275.

Roff, D. 1974b. The analysis of population model demonstrating the importance of dispersal in a heterogeneous environment. Oecologia 15:259–275.

Schlesselman, J. J. 1982. Case-control studies: design, conduct, analysis. New York: Oxford University Press.

Schindler, D. W., K. H. Mills, D. F. Malley, D. L. Findlay, J. A. Shearer, I. J. Davies, M. A. Turner, G. A. Linsley and D. R. Cruikshank. 1985. Long-term ecosystem stress: The effects of years of experimental acidification on a small lake. Science 228:1395–1401.

Shugart, H. H. 1984. A theory of forest dynamics. New York: Springer-Verlag.

Shugart, H. H. and D. C. West. 1977. Development of an Appalachian deciduous forest succession model and its application to assessment of the impact of chestnut blight. J. Environ. Manage. 5:161–179.

Slatkin, M. 1974. Competition and regional coexistence. Ecology 55:128–134.

Sommer, U. 1985. Seasonal succession of phytoplankton in Lake Constance. Bioscience 35:351–357.

Spiess, F. N., R. Hessler, G. Wilson and M. Weydert. 1987. Environmental effects of deep sea dredging. Scripps Institute of Oceanography.

Stewart-Oaten, A., W. Murdoch and K. Parker. 1986. Environmental impact assessment: "psuedoreplication" in time? Ecology 67:929–940.

Sullivan, P. F. and S. R. Carpenter. 1982. Evaluation of fourteen trophic state indices for phytoplankton of Indiana lakes and reservoirs. Environ. Pollut. (Series A) 17: 143–153.

Tessier, A. J. 1983. Coherence and horizontal movements of patches of *Holopedium gibberum* (Cladocera). Oecologia 60: 71–75.

Tessier, A. J. 1986. Comparative population regulation of two planktonic cladocera (*Holopedium gibberum* and *Daphnia catawba*). Ecology 67:285–302.

Threlkeld, S. T. 1979. The midsummer dynamics of two *Daphnia* species in Wintergreen Lake, Michigan. Ecology 60:165–179.

Tonn, W. M. 1985. Density compensation in *Umbra-Perca* fish assemblages of northern Wisconsin lakes. Ecology 66:415–429.

Tonn, W. M. and J. Magnuson. 1982. Patterns in the species composition and richness of fish assemblages in northern Wisconsin lakes. Ecology 63:1149–1166.

Vandermeer, J. H. 1973. On the regional stabilization of locally unstable predator-prey relationships. J. Theor. Biol. 41:161–170.

Vollenweider, R. A. 1968. Possibilities and limits of elementary models concerning the budget of substances in lakes. Arch. Hydrobiol. 66:1–36.

Wetzel, R. G. 1983. Limnology. Philadelphia: Saunders.

Yodzis, P. 1978. Competition for space and the structure of ecological communities. Berlin: Springer-Verlag.

Zeigler, B. P. 1976. The aggregation problem. *in*: Systems analysis and simulation in ecology, volume IV, ed. B.C. Patten, 299–311. New York: Academic Press.

Zeigler, B. P. 1977. Persistence and patchiness of predator-prey systems induced by discrete population exchange mechanisms. J. Theor. Biol. 67:687–713.

Zeigler, B. P. 1979. Multilevel multiformalism modeling: an ecosystem example. *in*: Theoretical systems ecology, ed. E. Halfon, 17–54. New York: Academic Press.

Part 5
Synthesis

CHAPTER 15

Goals of the Synthesis Groups

Stephen R. Carpenter

Synthesis groups were convened following the plenary papers and group discussions summarized in the preceding sections of this book. The purpose of the synthesis discussions was to address broad issues common to many of the subdisciplines of aquatic ecology. Synthesis groups sought to identify research needs and objectives for unraveling complex interactions in lake communities, and recommendations for overcoming both scientific and practical impediments to research on complex interactions.

Three replicate synthesis groups were convened, each with the same objectives. By this mechanism we hoped to diversify our recommendations and prevent domination of the synthesis process by a single, forcefully argued perspective. Each synthesis group had at least two representatives from each of the five topic-oriented discussion groups. Steve Bartell, Don Hall, Don McQueen, Bill Neill, Don Scavia, and Earl Werner served as discussion leaders and rapporteurs for the synthesis groups, and prepared written summaries of the synthesis discussions at the workshop. After the workshop it became clear that the most effective way to summarize these discussions was to combine the individual reports into a single document. Jim Kitchell undertook this task, with substantial input from seven coauthors.

We chose not to produce a consensus report, which might have suffered from sterility. Points of disagreement are evident, and will, we hope, provoke further examination and experimental tests. Not all authors agree with every point of the following chapter, but on debatable issues at least a majority of the authors subscribe to the views expressed in the text. This unconventional editorial practice has retained some of the flavor of the lively discussions which led to this chapter.

CHAPTER 16

Epistemology, Experiments, and Pragmatism

James F. Kitchell, Steven M. Bartell,
Stephen R. Carpenter, Donald J. Hall,
Donald J. McQueen, William E. Neill,
Donald Scavia, and Earl E. Werner

Introduction

In retrospect, what we now call complex interactions encompass several of the major advances in aquatic ecology. Some examples include the trophic-dynamic concept (Lindeman 1942), the multidimensional niche (Hutchinson 1957), size selective predation and the size efficiency hypothesis (Hrbacek et al. 1961; Hrbacek 1962; Brooks and Dodson 1965), the keystone predator concept (Paine 1966), and optimal foraging theory (Werner 1977). Recent advances offering similar benefit include the microbial loop (Riemann and Sondergaard 1986; Scavia and Fahnenstiel this volume; Porter et al. this volume), the ontogenic niche (Werner and Gilliam 1984; Crowder et al. chapter 10; Stein et al. chapter 11), chemical induction of antipredator morphological, behavioral and life history traits (Havel 1987), behavioral responses to predation (Kerfoot and Sih 1987) and the trophic cascade argument (Carpenter et al. 1985). The preceding chapters offer many specific examples of complex interactions in aquatic communities. Others will certainly appear in the near future. Chapters 10-14 develop both the general state of current understanding and the specific priorities for future work in this area of ecological research.

The goal of this chapter is to facilitate and accelerate the development of as yet unknown new ideas by offering a synthesis of current knowns and our best guesses about potentially productive directions for future work. Reasoning from theory and collective experience, we offer some epistemological perspectives with regard to studies of complex interactions. In addition, we make some specific suggestions about the institutional, logistic and financial arrangements that can enhance the creative potential of research activities.

Why Study Complex Interactions?

One respondent to the workshop invitation asked, "Why complex interactions? I'm still trying to figure out the simple ones!" The point was well taken. There are at least two major reasons why we cannot forego work on complex problems until all of the simple ones are resolved.

The primary reason for work on complex problems is that many of the specific questions may simply pale to insignificance (or be incidentally solved) as we learn more of the general mechanisms that structure aquatic communities and regulate their function. In other words, the big picture will emerge before each of the numbered spaces is filled. As argued from the perspective of hierarchy theory (Allen and Starr 1982; Allen et al. 1984; O'Neill et al. 1986; Frost et al. chapter 14), a more effective concept of scale allows understanding of what can and what cannot be resolved.

We have neither the time nor the talent to pursue our understanding of complex ecological interactions exclusively through reductionism. New insights will best, if not solely, derive from an effective hybridization of appropriately scaled concepts and methods.

The second reason for work on complex questions is the urgency of application-aquatic ecology must be closely allied with the management of both water quality and fisheries. That is both a desirable result of basic research and a societal obligation of the National Science Foundation. Success must be viewed in the larger, ultimate context of the success of ecology as a discipline. We might well profit from the example of molecular biology, where disciplinary proximity to applications in medicine has allowed a valuable synergism in generating support for and productivity of basic research. Our proximity to environmental issues and aquatic resource management offers similar potential. Redfield and Flanagan's message (Preface, this volume) is clear-be creative, collaborative, and do not avoid applied issues.

Temporal Patterns and Interaction Strength

Analyses summarized in this volume argue that a major effort should focus on the relationship between temporal and/or spatial pattern and the strength of interactions in lake communities (Bartell et al. chapter 7). As is apparent in the deliberations of the food web group (Crowder et al. chapter 10) and the habitat interactions group (Lodge et al. chapter 12), in the results of mesocosm experiments (Neill chapter 3) and in the larger context of whole lakes viewed on a historical scale (Carpenter chapter 8), surprising and apparently unpredictable responses arise from temporal and spatial heterogeneity in the existence or strength of key interactions.

The perspective of Paine's strong interactions (Paine 1981) and a recent review comparing terrestrial and aquatic community responses to predation

(Murdoch and Bence 1987) show that some of the most important linkages result in local extinctions. It follows that what cannot be seen in an initial, descriptive study may turn out to be the most important components of a set of complex interactions. It follows, too, that the apparent variability of populations and communities serves as evidence that linkage strength is variable.

The large and growing number of surprising responses of natural systems to experimental manipulation (Walters 1986) suggests that the equilibrium assumptions of conventional population and community ecology and food web theory may be misleading, if not wholly inappropriate (Crowder et al. chapter 10). While it is difficult to design for surprises, it is apparent that confining an experimental treatment series to the range of observed variation within a given system may not reveal the major mechanisms responsible for the current variation. The factors that determine community composition are not necessarily those that regulate interspecific interactions. The most informative surprises usually express change of at least fivefold to tenfold in response variables such as organism size, survivorship, percent composition of an initially dominant species, etc. Thus, a rule of thumb for experimentalists searching for indicators of linkage strength is to plan for at least a 10X range in treatment conditions and to be prepared for surprises in the results.

Surprising responses from manipulative experiments may result from scale dependence of system response in microcosms, mesocosms, etc. Miniaturization may retain a representative physico-chemical context for studying complex interactions; however, the organisms of interest cannot be rescaled. This may lead to amplification of the interactions (Neill chapter 3) and a subsequent misinterpretation of their relative importance in larger systems. This scaling problem suggests that examination of complex interactions through manipulative experiments be performed in a series of different sized enclosures in order to develop scaling rules for extrapolation to lakes and reservoirs.

Results measured across order-of-magnitude treatment ranges may also be of broader theoretical importance. Normally well-behaved, deterministic dynamic systems can exhibit aperiodic, unpredictable behavior when pushed (e.g., through manipulation) to specialized regions of parameter space (May and Oster 1976, Gleick 1987). Environmental, ecological, or energetic constraints may decrease the likelihood of naturally functioning systems entering these chaotic parameter regions. Careful experimental design integrated with appropriate models may provide a powerful combination for further elaborating complex interactions in communities and testing important theoretical concepts.

Ecological systems often manifest multiple causality (Hilborn and Stearns 1982). The same phenomenon (e.g., change in mean size) can result from different mechanisms (e.g., size selective mortality or change in resource supply). Future work on complex interactions will reveal many

more cases of multiple causality. Analyses of direct and indirect pathways
may effect a fusion of population and ecosystem paradigms in the arena
of aquatic community ecology. For example, the large-scale perspective
of biogeochemical budgets offers relatively poor predictability of community composition, as does the species- level resolution of population
ecology. Yet biogeochemical processes such as zooplankton excretion,
the function of microbial loops, and littoral and benthic detrital processing
strongly influence nutrient supply rate ratios to phytoplankton (Lodge et
al. chapter 12). Population interactions-predation and competition-may
then regulate composition of the assemblage. Each of these processes
typically operates on different spatial and temporal scales; budgets on the
annual and whole lake scale for biomass of entire trophic levels, while
the population processes include seasonal to instantaneous time scales
operating at the level of habitats, species, and life history stages. Reports
of the topic groups (Chapters 10-14) identify a diversity of major research
questions. It is apparent, however, that scale issues appear in each of
many perspectives and therefore merit the attention of major research
effort.

Hierarchical perspectives that help develop some general rules or
guidelines for experimental work on multiple time-space scales are sorely
needed. Because lakes offer such a wide range of conditions, they offer
opportunities to develop concrete examples of scale dependent phenomena
and applications of hierarchy theory. The deliberations of Frost et al.
(chapter 14) offer some general guidelines, but the specific examples of
conjecture and testing remain to be developed by creative researchers.

Ecological applications of hierarchy theory (O'Neill et al. 1986) draw
heavily on terrestrial examples and the results of incisive experimental
studies in the rocky intertidal (Paine 1980). Both are essentially two-di-
mensional and long-term relative to the dynamics of a planktonic community. One summer of phytoplankton turnover is analogous to thousands
of years of forest dynamics. Thus, the multidimensional problems of lakes
and the range of important temporal scales that operate in aquatic communities demand special and specific attention. The potential range of
scales is clearly demonstrated in the contrasts of the as yet immeasurably
rapid rates of the microbial loop (Porter et al. chapter 13), long-term observational studies (Mills and Forney chapter 2), and the view of centuries
hybridized from empiricism and simulation by Carpenter (chapter 8).

Recruitment Research: An Example, Opportunity, and Imperative

One of the major themes in this volume is the evidence of the linkage of
fishes to microbes and phytoplankton. As clearly argued by Persson et
al. (chapter 4), limnology has paid little attention to the role of fishes in

aquatic communities. That oversight is now obviously important. As an example of the need for further work on food web linkages and their role in complex interactions, we take one research theme, recruitment, and develop it more explicitly.

Understanding the causes of variability in recruitment of early life history stages of fishes has been identified as a key problem by both applied and basic research interests (May 1984, Rothschild 1986). Obviously, the predictive power of that understanding will enhance management capability of the fisheries profession. Much the same magnitude of variability is expressed by many marine and freshwater invertebrates. In the larger ecological context, Strong (1984) demonstrates that density vague regulation of populations and community interactions is evident in many aquatic and terrestrial systems.

A panel formed by the U.S. National Academy of Science identified the recruitment question as the foremost research priority in fish ecology (Rothschild 1986). Similarly, the National Oceanic and Atmospheric Administration and the Biological Oceanography Program of NSF have recently sponsored major research initiatives designed to help resolve the unknowns associated with the recruitment question.

One of the key questions identified through the Complex Interactions Workshop is the recruitment problem, although in the context of complex interactions, recruitment issues are more often expressed through the complexity of effects arising from changes in trophic ontogeny (Werner and Gilliam 1984). Resolution of both cause and effect with regard to variation in recruitment of fishes would greatly advance our understanding of aquatic community ecology. The recruitment process is known to be limited by a nested hierarchy of abiotic and biotic constraints. Autecological constraints dictate the presence or absence of a species. Variable, stochastic weather effects play a major role in the production and relative survival of gametes and early life stages. Species interactions such as predation, competition, disease, and parasitism subsequently determine most of the mortality of a cohort within the first few months of life. Both density dependent and density independent components are involved and interact on multiple spatial and temporal scales. Although the majority of attention has focused on fishes, virtually any species that exhibits pulsed reproductive effort and highly variable survivorship in early life history stages can evoke similar ecological effects. Thus, the same principles of trophic ontogeny and size structured interactions apply to many planktonic taxa (such as *Chaoborus* and many copepod species), as well as to a diversity of littoral and benthic invertebrates (chapters 10-12).

As the case for fishes is best known, their example will serve to illustrate the imperative for research. Year class strength in fishes is highly variable-a basic and oft-confirmed observation. Much of what we know about recruitment variation has derived from studies of fish populations manipulated through exploitation. As developed in a subsequent section on epistemology and experimental approach, big treatment effects expressed in recruitment responses have been a major source of insight.

The consequences of variable recruitment can be transmitted over long time scales (many years) and expressed throughout aquatic food webs. For example, a strong year class of planktivorous fishes can alter the composition of the zooplankton (Crowder et al. chapter 10; Mills and Forney chapter 2; Persson et al. chapter 4). As a result of altered species composition and biomass of zooplankton, effects cascade at smaller scales through altered selectivity and grazing rates into the phytoplankton and the microbial loop (Riemann and Sondergaard 1986; Kitchell and Carpenter 1988; Stockner and Porter chapter 5; Scavia and Fahnenstiel, chapter 6). As developed in several chapters of this book, fish effects are dependent upon and expressed at many levels of the community and vary in response to recruitment variation.

Variation in recruitment of a long-lived piscivorous species may evoke an even more extensive and long-lasting effect, but with opposite manifestations at each successively lower trophic level. The storage effect (Warner and Chesson 1985) of a strong and successfully recruited year class of piscivores can last a decade or more and will continue to be expressed in the composition and function of the aquatic community. This level of regulation and feedback is expressed in the opposite way at each lower trophic level, can completely alter planktonic community structure, and may account for up to 50% of the variance in primary production (Carpenter and Kitchell 1987; Carpenter et al. 1987). The key component in these cases is the cause of recruitment variation. The manifestations of variable recruitment afford an opportunity to examine amplification or attenuation of ecological signals through networks of populations connected by competitive or predator-prey interactions that vary through space and time.

Virtually every plausible cause for recruitment variation has been advanced and championed (May 1984). The arguments range across density independent mechanisms (e.g., storms, asynchrony in the seasonal plankton bloom relative to early life stages), simple starvation, strong competition, predation, and all possible interactions among the above (Rothschild 1986). Each is credible and none are unequivocally tested. Depending on the species in question, some mechanisms may be more important than others. Generality may emerge from the many research efforts currently directed to this issue but community ecology can neither wait for the answers nor do without them. In fact, we must view recruitment variation in the community context if we are to make progress. Variable recruitment is both the effect of species interactions and the cause of community changes. Understanding the causes, effects, and feedbacks of recruitment variation is a critical nexus of understanding complex interactions in aquatic systems.

Lake systems offer a unique opportunity for developing the requisite research on this issue. They are, in general, less subject to the confounding advective effects of marine systems. They offer relatively discrete and, therefore, quasireplicate systems for development of independent tests. Diversity at the level of species, life histories, and trophic guilds is variable among lakes-but usually in predictable ways (Tonn and Magnuson 1982)-and offers a gradient of potential interactions. Similarly, the range of morphometry, productivity and sizes offered in lake systems allows development and testing of ideas pertinent to the questions of habitat interactions, resource limitation, and scale.

Opportunities for natural experiments abound. Strong or weak year classes are often correlated within a regional setting; thus, the opportunity for independent replication is readily available. Controlled experimentation on recruitment can be accomplished through cooperation with fisheries management agencies. Many lakes are regularly manipulated by stocking policy and other large-scale fisheries management practices designed to alter specific fish populations and/or composition of the fish community. Exploitation regulations can effect substantial changes in the size and species composition of the fish assemblage. Piscicides are regularly used to remove undesirable fishes. The resultant communities and their interactions offer a full range of treatment conditions from zero to high density fish populations, and a set of intermediates that would allow evaluation of colonization responses, founder effects and understanding of the shape of response curves as community development proceeds.

In small systems, manipulation at the species or trophic level are easily achieved, while in large systems (e.g., the Laurentian Great Lakes) analogous manipulations are underway and can serve as a basis for tests of scale effects (Kitchell and Carpenter 1987; Scavia and Fahnenstiel chapter 6). Natural experiments such as winter- or summer-kills occur regularly in some systems and intermittently in others. Recolonization and the development of a fish species assemblage occurs through natural immigration processes and/or through the action of fisheries agencies.

Thus, the ontogenically changing role of fishes as predators and competitors is amenable to experimental analysis at temporal and spatial scales pertinent to complex interactions in natural communities. A diversity of opportunities exists for collaboration with management agencies in developing a suite of manipulative studies that can reduce the costs of large-scale experimentation. The resultant convergence of basic and applied interests is an obvious advantage.

The Case for Opportunism

Planning exercises such as this workshop should be complemented by arguments on behalf of serendipitous opportunity. Community ecology has long taken advantage of natural experiments and an opportunistic response to major natural catastrophes in order to develop understanding or generate hypotheses complementing those derived from controlled experimentation. Manipulations performed by lake managers offer similar opportunities.

Many classes of natural experiments are available in lake systems. Winter-kill lakes have been used as a means for sorting among the effects of autecological constraints and species interactions (Tonn and Magnuson 1982). Some of the most dramatic insights derived from the history of aquatic community ecology stemmed from the introduction of exotic species. Some examples include alewife in coastal ponds (Brooks and Dodson 1965), peacock bass in Gatun Lake (Zaret and Paine 1973), Nile perch in Lake Victoria (Balon and Burton 1986), and the sequence of sea lamprey, alewife and Pacific salmon in the Great Lakes (Kitchell and Crowder 1986). Other invasions such as those of the Eurasian water milfoil, *Hydrilla*, water hyacinth, rusty crayfish, and grass carp have had equally potent and instructive effects in developing our understanding of littoral zone communities (Carpenter and Lodge 1986). At the smaller scale, experimental approaches substantially enhance understanding and predictive power when dominant or rare members of an extant community are manipulated (Paine 1980, Bergquist and Carpenter 1986, Neill chapter 3, Persson et al. chapter 4).

Most invasions or introductions have had little ecological consequence (Diamond and Case 1986); however, the spectacular effects of those that have been successful demand attention to the future and will challenge the very best of ecological theory (Balon and Burton 1986). Ecological principles have been successfully employed to forecast the outcome of some introductions (Kitchell and Crowder 1986), but most are poorly anticipated and fully expressed before ecologists can offer either advice to management or contributions based on firm understanding.

From the lessons of history, we can be confident that exotic species will continue to appear in lake systems. We can also see that a planned rather than reactive approach will recognize these perhaps unwanted introductions as an opportunity to learn. Clearly, our understanding of rules for community composition have been augmented by the lessons of previous invasions. It follows that invasions to come may be equally if not more instructive. Lake systems offer the obvious value of replicates in time and space for this next in a series of potentially instructive mishaps.

Epistemology and the Experimental Approach

No single experimental or analytical approach seems uniquely appropriate to the evaluation of complex interactions. There is, however, some guidance from the principles of hierarchy theory. Given the number and kinds of unknowns and the frequency of unexpected results that derive from

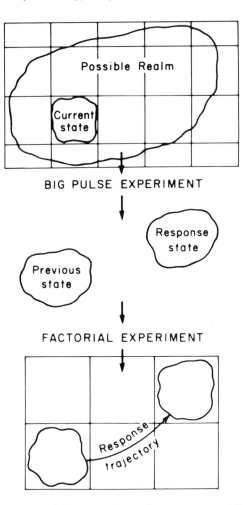

FIGURE 16.1. A conceptual diagram of experimental protocol for evaluating complex interactions. The current state and maximum response of a community are first determined by a big treatment experiment. Rate of response, shape of the response curve, and interactions of system components are subsequently determined via factorial experiments. Only two of potentially many system state variables are depicted.

research on lake communities, it follows that a sequential or hierarchical treatment protocol would accelerate our rate of discovery.

Lessons learned from the introductions of exotic species (see previous section) or large-scale biogeochemical perturbations (Schindler and Fee 1974; Schindler et al. 1985) clearly demonstrate big effects from big treatments. These lead to an understanding of the dimension of potential community response and allow inferences about the relative importance of specific mechanisms as the system changes state. Thus, a first approach to gaining understanding would involve a plan for applying big treatment effects and anticipating big responses. The major merit and goal of this approach is to dimension the limits of possible community responses as a first estimate of the compensatory capacity of the system under study.

In Chapter 10, arguments are advanced against working outside the realm of observed variation. One of the perspectives of a hierarchical view is that system structure can best be understood by defining system limits (O'Neill et al. 1986). We argue that some of our major conceptual advances (e.g., the keystone predator concept) derived from an approach based on big treatment manipulations, that the lessons of exotic species include many unexpected responses, and that even well-planned experiments evoke most informative surprises (Neill chapter 3). We should plan for and initiate this type of manipulation as a first step.

We recognize, however, that the big perturbation approach offers less understanding of the transient behavior of complex interactions as expressed at the temporal and spatial scales of natural variability. Toward an understanding of those questions, we suggest a factorial or gradient design within the range of variation exhibited under large perturbations. This would allow sufficient replication to elucidate mechanisms, clarify the components of variability, quantify directional change, and detect nonlinearities. Thus, the mechanistic understanding required for adequate prediction and testing would be best derived by first conducting a big manipulation experiment designed to dimension the response capacity of the community, followed by factorial or gradient experiments focusing on certain components of the reduced matrix of possibilities (Fig. 16.1).

Although experimentalists (acting on the advice of good statisticians) might choose to develop a factorial design as a first approach, the lessons of history suggest that a big treatment experiment followed by gradational treatments will maximize our progress. That evidence lies in the many mesocosm and pond studies where the major contribution of the research was evoked in only a small fraction of the treatment units (Hall et al. 1970, Neill chapter 3). Judicious monitoring of community structure following natural or planned manipulations represents an alternative to the combined manipulation and gradient experiments previously outlined. Estimates of the variance associated with community components through space and time permit the development of neutral models (Caswell 1976) aimed at quantifying the relative importance of different ecological interactions (Harris and Griffiths 1987). The ability of these variously scaled models to explain portions of the monitored variance in community structure may reveal the scale dependence of interaction strength. Development, application, and analysis of these models may provide a means for identifying the scales at which various phenomena, identified and measured in the laboratory, apply in nature.

One of the workshop synthesis groups discussed experiences with unexpected results that arose during their research programs. Their consensus was that major conceptual insight occurs as a revelation. The template for creativity is usually opened by conflict, paradox, or contradiction. The catalysis of innovation is generally unpredictable, but usually associated with an alternative offered by abandoning the constraints of conventional wisdom, by the insights of a colleague trained to a different world view, or by an accident. While it remains impossible to plan for creative insights, we can recognize that thinking at a different scale and interaction with scientists of different traditions are common correlates. Our recommendations at the close of this chapter emphasize development of the kinds of opportunities and institutional support that may provide the requisites of creativity.

Many of our most informative surprises have derived from unexpected perturbations in complex systems (Walters 1986). These revealed the constraints of response potential (O'Neill et al. 1986). We suggest that assessing the extreme scenarios of community response would be a most appropriate first step. That helps define the determinants of community structure. Analysis of variation around the mean condition can follow. That provides understanding of the regulators of community structure. Given the finite lifetime of most research grants and the low funding probability of a renewal to clean up the details, we argue that our understanding of response patterns will best derive from a bold initial step and a well-planned follow through. Progress may be further enhanced if this kind of protocol can be employed in an institutional setting that maximizes the prospect for conceptual insight and collegial interaction.

Fostering Creativity

The proliferation of pages in print has a negative effect on the kinds of creative endeavor required to better understand complex interactions. A logical consequence of the information explosion is that individual researchers will choose and/or be forced by time constraints into increasing specialization. This tendency is counter to the kinds of expansive thinking that may help resolve large, complex problems. In fact, specialization in ecology tends to promote refinement of the status quo, which is antithetical to the development of fundamentally new ideas-the charge of the National Science Foundation!

How can this contradiction be resolved? We argue that efforts to promote interdisciplinary collaboration and training are required. Research initiatives should maximize learning rates through pairings of theorist and experimentalist, molecular microbiologist and plankton ecologist, fisheries biologist and limnologist, etc. In a similar way, studies that combine two or more scales of investigation should prove fruitful. For example, the patchiness of planktonic and littoral habitats could profit from the view

of those who sample on the scale of minutes and meters if properly juxtaposed with the view held by those whose perspective is of whole lakes
and years to decades. This hypothetical combination can take the real
world shape of collaboration among algal ecologist, fish ecologist and pa-
leoecologist.

The greatest rates of progress on complex interactions will arise from
polythetic approaches. The various approaches to limnological research
differ in the space and time scales which they address as well as in their
degree of realism, detail, and complexity (Fig. 16.2). All models employ
simplification to achieve insight, and so are low in realism, detail, and
complexity. However, models offer great flexibility with respect to time
scale. Laboratory microcosms achieve somewhat greater realism than
models, but generally operate on restricted time scales of hours to months.
A somewhat longer range of time scales is addressed by field mesocosms,
which offer relatively more realism, detail, and complexity. Whole-lake
results are the standard of realism, detail, and complexity to which all
other approaches must be compared. However, field experiments at large
spatial and temporal scales are rather rare (Strayer et al. 1986). Paleolimnology is the only consistent source of integrative data at time scales of
decades to centuries, but such data span a great range of levels of realism,
detail, and complexity (Binford et al. 1983).

Aquatic ecologists are asked by society to provide information for management decisions that usually focus on time scales of years to decades.
Models, paleolimnology, and whole-lake studies are the only approaches
that directly address these time scales. However, a number of constraints

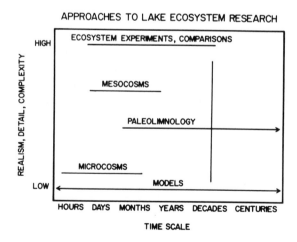

FIGURE 16.2. Relationships among possible approaches to analyzing lake systems
with respect to time scale on which each approach can be applied, and the degree
of realism, detail, and complexity of each approach.

on whole-lake studies also dictate important roles for work in microcosms and mesocosms. Multiple causality is ubiquitous, and the methods for separating and quantifying contrasting causal mechanisms dictate small-scale, reductionistic studies (Hilborn and Stearns 1982). Yet, small-scale studies alone are not sufficient for forecasting behavior of larger systems, because small-scale studies cannot fully determine which causal pathways will predominate at larger scales. One of the great challenges of ecology is to understand how information from models, small-scale experiments, and paleoecology can be translated into inferences and predictions about long-term lake dynamics. Integrative studies combining several approaches and scales will be needed to resolve this problem.

"In My Lake..."

We begin this section by refusing to concede that each lake is different from all others. That may seem unnecessary to those who have not attended a national limnological meeting but the phrases "In my lake..." and "But in my lake..." should sound disturbingly familiar to those who have. From certain research perspectives, the individuality of lakes is an interesting property and an asset (cf. Persson et al. chapter 4). We feel, however, that studies of complex interactions will advance most rapidly if we can develop ways to understand basic mechanisms before testing their generality in the vast range of lake types. In fact, we argue that the test for site effects, while an important consideration, is subordinate to derivation of a mechanistic understanding.

Methods must be developed that permit extrapolation of interaction strength across systems of different scale. No rigorous criteria for valid comparison across differently sized systems have been established. Investigators invoke names (i.e., microcosm, mesocosm, pond, lake) that carry implications of structure, complexity, and scale; nevertheless, these names remain attached to fuzzy concepts that continue to frustrate the development of theory and promulgate the uniqueness of individual study sites.

We have attached high priority to the collaborative, interdisciplinary approach. Given the dispersion of aquatic ecologists, we conclude that research initiatives will be most effective if specialists can be aggregated by common interest at a subset of common sites. We envision development of facilities that will focus people and research problems, thereby reducing the unnecessary duplication of analytical facilities and monitoring effort that accompanies limnological work on each of many lakes. Thus, the overhead of research enterprise can be minimized and the creativity of scientists more effectively focused on important research questions.

Based on discussions during the workshop, at least four general kinds of sites seem necessary: arctic or alpine lakes, natural north temperate lakes, southern reservoirs, and tropical lakes. This ensemble would offer the full range of physical and latitudinal gradients of interest. Each site should include:

1. Multiple lakes, some monitored, others available for large-scale manipulation.
2. Administrative infrastructure and permanent staff to coordinate and maintain the facility.
3. Capacity for smaller-scale manipulative studies as might be conducted in pond systems, large limnocorrals, or mesocosms.
4. Laboratory facilities, housing, and appropriate local support staff as necessary to accommodate independent investigators.

Coordination among sites would permit tests of hypotheses across trophic or geographic gradients. An example would compare the magnitude of food web effects in natural oligotrophic lakes with those derived from similar manipulations in eutrophic reservoirs (see Crowder et al. chapter 10). Results of these kinds of comparative tests would serve as a basis for developing more rigorous tests for site effects and create some perspective on the merits of the argument that "In my lake..."

In much the same fashion as oceanographers share cruise opportunities, independent investigators interested in a particular aspect of complex interactions would be encouraged to pursue their expertise under the umbrella of ongoing or planned large-scale manipulations. Again, as in the case of shipboard collegiality, the synergism of informal interactions among juxtaposed scientists provides opportunities for new ideas and creative exchange that minimize the inefficiencies of isolated competition and the constraints of traditional wisdom.

These arguments seem familiar to those of us who have pleaded the case for a new colleague or support to a field station. Our goals differ in that the focus of these facilities would be ongoing and intended large-scale manipulations. We hold that a system of experimental lake sites is extremely valuable to aquatic ecology and deserving of the same kind of institutional support that has been provided for research vessels, high energy physics installations and biotechnology centers. More importantly, such a system is requisite for most efficient and rapid progress in developing our understanding of aquatic communities, their variability, and their potential as public resources.

Additional Opportunities

The general goals above can also be facilitated by a number of more modest programs.

1. Provide more flexibility in the grant process.
   a. Fund collaborative efforts appropriate to the disciplinary diversity of the problem.

b. Fund studies for longer periods than the traditional two or three years when such longer time scales are appropriate to the problem.

c. Provide extensions of up to 12 months which would allow time for data interpretation and development of renewal requests in phase with the annual cycle of field research.

2. Provide more support for interaction and collaboration.

a. Provide grant supplements to allow travel for consultantships and collaborations.

b. Support general planning workshops (such as the Notre Dame workshop) at about 5-year intervals.

c. Support annual series of special topic conferences in aquatic ecology similar to the Gordon Conferences.

3. Create opportunities for broader training.

a. Support training workshops on new techniques or approaches (e.g., microbial methods, phytoplankton taxonomy, applications of new theory, etc.).

b. Provide a greater number and diversity of fellowships for graduate students (e.g., to support travel and yearlystipends to learn new techniques and perspectives at a different institution; to support summer work learning new techniques; to support the development of curricula that broaden students' training).

c. Provide postdoctoral and midcareer training for scientists seeking to learn a new subdiscipline and/or collaborate with workers in another discipline.

Summary

Complex interactions involve multiple causal pathways plus multiple spatial and temporal scales. Any complex interaction may have seasonal, biogeochemical, predator-prey, behavioral, and/or evolutionary components. Because of their multifarious nature, complex interactions are not dealt with effectively by any one of the established world views (epistemes) of aquatic ecology. For example, the taxocene approach (phytoplankton communities, fish communities, etc.) is inadequate because it neglects trophic structure, while the trophic level approach is often too aggregated to cope with variable life histories and most interspecific interactions. Consequently, progress on complex interactions will require new, or at least synergistic, combinations of the well-established epistemes. Complex interactions also require polythetic approaches. We argue that the most rapid progress will come from sequential research designs that proceed from strong manipulations at large scales to finer grained experimental programs designed to elucidate and compare individual mechanisms. In some cases, broadly-trained investigators and/or interactive teams will be needed to accomplish such research plans.

Funding effective work on complex interactions poses a significant

challenge. Intense competition for research support engenders conserv-
atism which favors work within the boundaries of conventional wisdom,
rather than the novel juxtapositions of perspectives, disciplines, and
methodologies that are needed to study complex interactions. Specific
suggestions for new kinds of funding and greater flexibility in the granting
process were contributed by many workshop participants and are summarized herein.

Finally, we must find ways of training researchers to recognize and
account for the effects of scale. Many forces, including intra-disciplinary
competition and the explosive growth of the literature, contribute to specialization. The resultant factionalization leads to an emphasis of small
problems, diminishing ecology's standing relative to the disciplines with
which it must compete for funds.Innovative programs are needed that
teach scientists to seek appropriately scaled approaches and foster the
receptiveness to other perspectives essential for progress on the most important basic and applied questions that face aquatic ecology.

Acknowledgements. We thank the workshop participants for their willingness to offer and argue the ideas developed herein. We also thank Garth
Redfield and Patrick Flanagan for their initiative and support throughout
the development of this workshop. J.F.K. thanks Mary Smith for ex-
tracting the manuscript from a middle number system of computer files
and marginalia.

References

Allen, T. F. H. and T. B. Starr. 1982. Hierarchy: Perspectives for ecological complexity. Chicago: University of Chicago Press.
Allen, T. F. H., R. V. O'Neill, and T. W. Hoekstra. 1984. Interlevel relations in
ecological research and management: Some working principles from hierarchy
theory. USDA Forest Service General Technical Report RM-110, Rocky
Mountain Forest and Range Experiment Station, Fort Collins, Colorado, 11 pp.
Balon, E. K., and M. N. Burton. 1986. Introduction of alien species or why scientific advice is not heeded. Env. Biol. Fish. 16:225–230.
Bergquist, A. M., and S. R. Carpenter. 1986.Limnetic herbivory: effects on phytoplankton populations and primary production. Ecology 67:1351–1360.
Binford, M., E.Deevey, and T. Crisman. 1983. Paleolimnology: An historical per-
spective on lacustrine ecosystems. Ann. Rev. Ecol. Syst. 14:255–286.
Brooks, J. L., and S. I. Dodson. 1965. Predation, body size, and composition of
plankton. Science 150:28–35.
Carpenter, S. R., J. F. Kitchell, and J. R. Hodgson. 1985. Cascading trophic in-
teractions and lake productivity. BioScience 35:634–639.
Carpenter, S. R., and J. F. Kitchell. 1987. The temporal scale of limnetic primary
production. Am. Nat. 129:417–433.
Carpenter, S. R., J. F. Kitchell, J. R. Hodgson, P. A. Cochran, J. J. Elser, M.
M. Elser, D. M. Lodge, D. Kretchmer, X. He, and C. N. von Ende. 1987.

Regulation of lake primary productivity by food web structure. Ecology 68: 1863–1876.

Carpenter, S. R. and D. M. Lodge. 1986. Effect of submersed macrophytes on ecosystem processes. Aquat. Bot. 26:341–370.

Caswell, H. 1976. Community structure: a neutral model analysis. Ecol. Monogr. 46:327–354.

Diamond, J., and T. J. Case. 1986. Ecological Communities. New York: Harper and Row.

Gleick, J. 1987. Chaos. New York: Doubleday.

Harris, G. and F. B. Griffiths. 1987. On means and variances in aquatic food chains and recruitment to the fisheries. Freshwat. Biol. 17:381–386.

Havel, J. E. 1986. Predator-induced defenses: A review. *in*: Predation: Direct and Indirect Impacts on Aquatic Communities, ed. W. C. Kerfoot and A. Sih. Hanover: University Press of New England.

Hilborn, R, and S. C. Stearns. 1982. On inference and evolutionary biology: the problem of multiple causes. Biotheoretica 31:145–164.

Hrbacek, J. 1962. Species composition and the amount of zooplankton in relation to the fish stock. Rozpr. Cesk. Acad. Ved Rada Mat. Prir. Ved. 72:1–116.

Hrbacek, J., M. Dvorakova, V. Korinek, and L. Prochazkova. 1961. Demonstration of the effect of the fish stock on the species composition of zooplankton and the intensity of metabolism of the whole plankton assemblage. Verh. Int. Ver. Theoret. Angew. Limnol. 14:192–195.

Hutchinson, G. E. 1957. Concluding remarks. Cold Spring Harbor Symp. Quant. Biol. 22:415–427.

Kerfoot, W. C., and A. Sih. 1987. Predation: Direct and Indirect Impacts on Aquatic Communities. Hanover: University Press of New England.

Kitchell, J. F., and S. R. Carpenter. 1987. Piscivores, planktivores, fossils and phorbins. *in*: Predation: Direct and Indirect Impacts on Aquatic Communities, ed. W. C. Kerfoot and A. Sih. Hanover: University Press of New England.

Kitchell, J. F., and S. R. Carpenter. 1988. Food web manipulation in experimental lakes. Verh. Internat. Verein. Limnol. 23:351–358.

Kitchell, J. F., and L. B. Crowder. 1986. Predator-prey systems in Lake Michigan: model predictions and recent dynamics. Environ. Biol. Fishes 16:205–211.

Lindeman, R. L. 1942. The trophic-dynamic aspect of ecology. Ecology 23:399–418.

May, R. M., ed. 1984. Exploitation of marine communities. Berlin: Springer-Verlag.

May, R. M. and G. F. Oster. 1976. Bifurcations and dynamic complexity in simple ecological models. Am. Nat. 110:573–599.

O'Neill, R. V., D. L. DeAngelis, J. B. Waide, and T. F. H. Allen. 1986. A Hierarchical Concept of Ecosystems. Princeton: Princeton Univ. Press.

Paine, R. T. 1966. Food web complexity and species diversity. Am. Nat. 100:65–75.

Paine, R. T. 1980. Food webs, linkage interaction strength, and community infrastructure. J. Anim. Ecol. 49:667–685.

Riemann, B., and B. Sondergaard. 1986. Carbon dynamics of eutrophic, temperate lakes. Amsterdam: Elsevier Scientific Publishers.

Rothschild, B. J. 1986. Dynamics of marine fish populations. Cambridge: Harvard University Press.

Schindler, D. W. and E. J. Fee. 1974. Experimental Lakes Area: Whole-lake experiments in eutrophication. J. Fish. Res. Bd. Canada 31:937–953.

Schindler, D. W., K. H. Mills, D. F. Malley, D. L. Findlay, J. A. Shearer, I. J. Davies, M. A. Turner, G. A. Linsey, and D. R. Cruikshank. 1985. Long-term ecosystem stress: The effects of years of experimental acidification on a small lake. Science 228:1395–1401.

Strayer, D., and J. S. Glitzenstein, C. G. Jones, J. Kolasa, G. E. Likens, M. J. McDonnell, G. G. Parker, and S. T. A. Pickett. 1986. Long-term ecological studies: An illustrated account of their design, operation, and importance to ecology. Pub. No. 2 of the Inst. of Ecosystem Studies, NY Bot. Garden. 38 pp.

Strong, D. R. 1986. Density vagueness: abiding the variance in the demography of real populations. in: Ecological Communities, ed. J. Diamond and T. J. Case, 257–268. New York: Harper and Row.

Tonn, W. M., and J. J. Magnuson. 1982. Patterns in the species composition and richness of fish assemblages in northern Wisconsin lakes. Ecology 63:1149–1166.

Walters, C. 1986. Adaptive management of renewable resources. New York: MacMillan.

Warner, R. R., and P. L. Chesson. 1985. Coexistence mediated by recruitment fluctuations: a field guide to the storage effect. Am. Nat. 125:769–787.

Werner, E. E. 1977. Species packing and niche complementarity in three sunfishes. Am. Nat. 111:553–578.

Werner, E. E., and J. F. Gilliam. 1984. The ontogenetic niche and species interactions in size-structured populations. Ann. Rev. Ecol. Syst. 15:393–425.

Zaret, T.M., and R.T. Paine. 1973. Species introduction in a tropical lake. Science 182:449–455.

Index